How to Build a Habitable Planet

Wallace S. Broecker

This book is distributed through the
Lamont-Doherty Geological Observatory
of Columbia University
Address all inquiries to:
ELDIGIO PRESS
LDGO BOX #2
Palisades, New York 10964

Printed in the United States of America

Acknowledgements

The layout and typesetting for this book was done by
Computer Artistry
Tucson, Arizona

The cover design was done by
Scott Broecker

Foreword

This book is an outgrowth of an undergraduate course I taught for students from Columbia College and Barnard College. My approach was to trace the development of the Earth from its roots in the Big Bang to its future in man's hands. At each stage I depict the important observational evidence and attempt to show how it has been linked together into hypotheses. I attempt to show that while in some cases this evidence constrains us to a widely accepted single hypothesis, in others the evidence is incomplete and permits a range of competing explanations. It was my hope that in this way I could bring my students to see that science is far from static. Rather, it is a series of ongoing studies where all the evidence, assumptions, and hypotheses are continually being reexamined and where new information is regularly being added.

Many readers will surely ask why this book was not published through the the usual channels. Beyond my liking for the unconventional, there is an important financial reason. For each book sold $2.50 will be returned to the Department of Geological Sciences as repayment of typing and drafting costs. After a year of negotiations with various publishers I found this to be the only way I could recover these costs. Thanks to computer layout schemes and laser printers, it is now possible to circumvent the very high overhead associated with conventional publishing.

I would like to thank Vicky Costello who patiently typed and retyped the many drafts of this manuscript, and Patty Catanzaro who drafted and redrafted the many figures. I also thank the ten or so scientists who read through the manuscript and made valuable suggestions. Finally, I thank the students in Geology 1011x during the years 1981 to 1985. They were the guinea pigs as well as the inspiration for this project.

To the late Paul Gast, who pioneered the trace-element approach to planetary geochemistry. From the time he was assigned by Wheaton College as my sophomore "big brother," until his death in 1976, Paul was a friend and inspiration. He rescued me from becoming an actuary and turned my energies toward geochemistry. I owe much to him.

Contents

Whirlpool Galaxy

Chapter One

The Setting:

The Big Bang and Galaxy Formation

Earth is a minor member of a system of planets in orbit around a star we call the Sun. The Sun is one of about 100 billion stars that make up our Milky Way. The light from this myriad of stars allows observers in neighboring galaxies to define our galaxy's spiral form. The galaxy is the basic unit into which universe matter is subdivided.

Like its billions of fellow galaxies, ours is speeding out on the wings of a great explosion that gave birth to the universe. That these major pieces of the universe are flying away from each other is revealed by a shift toward red in the spectra of the light reaching us from distant galaxies. The close correlation between the magnitude of this shift and the distance of the galaxy tells us that about 15 billion years ago all the galaxies must have been in one place. The catastrophic beginning of the universe is heralded by a dull glow of background light. This glow is the remnant of the great flash that occurred when the debris from the explosion cooled to the point where the electrons could be captured into orbits around the hydrogen and helium nuclei. Contained in the galaxies lying within the range of our telescopes are about 100 billion billion stars. A sizable number of these stars are thought to have planetary systems. While planets with Earth's highly favorable environment are surely exceptions, it is difficult to believe that Earth is unique. Other equally suitable habitats for life must exist.

Introduction

Why are we here? Are others like us to be found elsewhere? Difficult questions indeed! Despite millennia of thought the answers available today are little more satisfying than those that were available to our distant ancestors. New insights have been more than counterbalanced by new puzzles. Religion tells us we were created by God. Science tells us we were created by chance. Yet neither theologians nor scientists are quite happy with these pat answers. Both probe for greater insight. Are we God's only charges? Has chance operated successfully elsewhere? Perhaps someday God will reveal himself or chance will be duplicated in the laboratory. In the meantime, humans will continue to puzzle and probe.

While we as yet have no way to know whether living beings who match or exceed our ability to appreciate, to reason, and to exploit reside elsewhere in the universe, science is providing some very valuable constraints on such speculations. Two approaches are being taken. One is geological. It has to do with the likelihood that habitats suitable to life have come into being elsewhere. The other is biological. It has to do with the likelihood that given a suitable environment, life will evolve. Although, to date, only a few steps have been taken along these long roads, important progress has been made. In this book I attempt to summarize the success of the geologic approach.

In trying to assess the likelihood that other settings suitable for intelligent life exist, we must first have an idea of how many planets and moons there are in the universe. This proves to be an awkward question,because to date even planets associated with our nearest stellar neighbors are undetectable. Since they do not glow, we cannot see them. They are too tiny to create shadows or to measurably perturb their host star's path. Newly developed sensors mounted on the soon-to-be-launched space telescope will, it is hoped, give us our first view of distant planets.

Thus, the only planets we are sure about are those circling our Sun. Excluding those objects that are less than 100 kilometers in diameter (which, as we shall see, are clearly incapable of spawning intelligent life), there are nine planets and 40 or so moons. These objects give every appearance of being the by-products of the Sun's formation. This leads most astronomers to believe that, at least for stars in the size class of the Sun, planets may be the rule rather than the exception. If this speculation is correct, then there may be as many or more planets as there are stars! The total would be a staggering 10^{20}* objects. To comprehend this number, think of groups with 100 billion objects each. There would be a billion such groups. One might say at this point there is no need to speculate further; with so many planets, there must be myriads of them on which conditions favorable to intelligent life exist. Before we jump to this conclusion, however, let us hold open the possibility that our Earth is so unusual that there is only one chance in 10^{21} of it being duplicated. In this case, there would be only one chance in ten that another Earth exists. While not promising to be able to provide any useful evidence regarding the actual magnitude of this probability, the evidence in the chapters that follow leads us to believe that Earth conditions are not so improbable as to make our planet unique. There must be hosts of planets that have the critical prerequisites.

Table 1-1. Shorthand system for expressing very large and very small numbers

1,000,000,000	1×10^{9}
1,000,000	1×10^{6}
1,000	1×10^{3}
1	1×10^{0}
0.001	1×10^{-3}
0.000001	1×10^{-6}

In order to assess the likelihood that Earth conditions are duplicated elsewhere in the universe, we must go back to the beginning. Even the very early events in the universe left their mark on the Earth.

*If you are not familiar with this notation, please study the examples in Table 1-1.

The Big Bang

Astronomers tell us that the universe as we know it began about 15 billion years ago with an explosion they call the "big bang." All the matter in the universe still rides forth on the wings of this blast. Speculations as to the nature of this cosmic event constitute the forefront of a field called cosmology. What went on before this explosion is a matter that must be left to philosophers. No record of prior events remains.

For astronomers to state that they know the age of our universe and its mode of origin is rather bold. Is this fanciful thinking or do they have firm evidence? Surprisingly, astronomers have made observations which most scientists accept as compelling support for the big-bang theory of universe origin. On a scale of reliability that goes from zero (idle speculation) to 10 (proven), this theory gets a 9!

Before presenting this evidence, let us consider a seeming paradox that confronted astronomers before the concept of an expanding universe was proposed. Stated simply, no one was able to explain the fact that the nighttime sky is dark. The black background between the stars seemed to demand either that the universe have a finite extent or that the light from the most distant stars be intercepted by dark matter in the voids of space. To understand this, one has only to envision a universe of infinite extent made up of luminous objects separated by empty voids. In such a universe no matter where we looked we would see the light from some distant star. The sky would be blindingly bright! The obvious alternative is that the universe is finite. In a finite universe, we could look between the stars into the black void beyond. Another possibility is that there are clouds of nonluminous matter floating in the voids between stars and that these clouds block the light from the more distant stars from our view.

The first alternative appeared unacceptable because in a universe of finite extent there would be nothing to hold the stars apart. The mutual star-to-star gravitational attraction would lead to an unbalanced pull toward the "middle" of the universe. It would be as if we secured a series of balls on a great three-dimensional latticework and then connected each ball to each of the other balls with a stretched rubber band. While the balls near the center of the latticework would be pulled more or less equally from all directions, those near the edge of the lattice would be pulled toward the inside. If by magic we could suddenly remove the latticework, leaving only the balls and stretched rubber bands, there would be a massive implosion as the balls streaked toward what had been the lattice's center. Only if the lattice were infinite in extent would nothing happen. In this case the pull on every ball would be exactly balanced. The universe has no latticework to hold the stars apart, yet there they are. Hence, the finite universe explanation for the dark sky must be rejected as inadequate.

The second explanation—that the light from very distant stars is intercepted by dark clouds of dust and gas along its path to the Earth—is also unacceptable. In this situation the light from stars at intermediate distances would also be affected. We should see a glow of scattered light similar to that in the night sky over a great city or from headlights approaching through fog. No such glow is seen! So this explanation also must be rejected.

This paradox was articulated by Heinrich Olbers in 1826. More than a hundred years passed before this cosmological puzzle was solved. In 1927 a Belgian astronomer, Georges Lemaître, proposed that the universe began with the explosion of a cosmic "egg." This clever concept neatly explained the long-standing paradox in that the force of the explosion prevents the gravitational pull from drawing the matter toward the center of the universe. It would be as if a bomb were to blow the balls on our lattice away, overpowering the pull of the rubber bands. In the absence of observational evidence, the Lemaître hypothesis would have received relatively little attention. Within two years of its publication, however, Edwin Hubble reported observations that turned the attention of the scientific world toward the concept of an expanding universe. Hubble reported a shift toward the red in the spectra of light reaching us from the stars in very distant galaxies. The simplest explanation for such a shift was that these distant galaxies were running away from ours at incredible speeds.

The Red Shift

The light coming from the Sun consists of a wide spectrum of frequencies. As these light rays pass into and out of raindrops, they are bent. Each frequency is bent at a slightly different angle separating the bundle of mixed light into a spectrum of individual color components. This spectrum is a rainbow. Each of these frequencies leaves a different imprint on the retina of our eyes. We see them as colors.

Physicists more than a hundred years ago learned to duplicate the rainbow effect by passing sunlight through a prism of glass. Light rays passing through a raindrop are bent according to frequency. As shown in Figure 1-1, the red light (that with the lowest frequency detectable by our eyes) is bent the least, and violet light (that with the highest frequency detectable by our eyes) is bent the most.

Astronomers have long used prisms (and more recently diffraction gratings) in their telescopes as a means of examining the color composition of the light from distant galaxies. They observed imperfections in these spectra; regardless of the light source, dark bands were always found to mar the otherwise smooth transition from red to orange, to yellow, to green, to blue, and to violet. Of great interest was the fact that these dark bands are produced by the absorption of certain frequencies of light by the halo gas surrounding the star that is producing the light. While transparent to some frequencies of light, this gas is partially opaque to other frequencies. When examined in detail, thousands of these bands become apparent. Most do not completely blacken the rainbow; rather, they produce a weakening of the intensity of the light at that frequency. This weakening is the result of partial absorption of the departing light by the star's "atmosphere." Some "packets" of light are captured; others slip through and reach the Earth.

Astronomers originally took an interest in these bands because they offered a means of making chemical analyses of the star's halo of gas and hence of the star itself. Each partially darkened line in the spectrum represented a single element. A packet of light can interact with an atom only if it has just the right energy

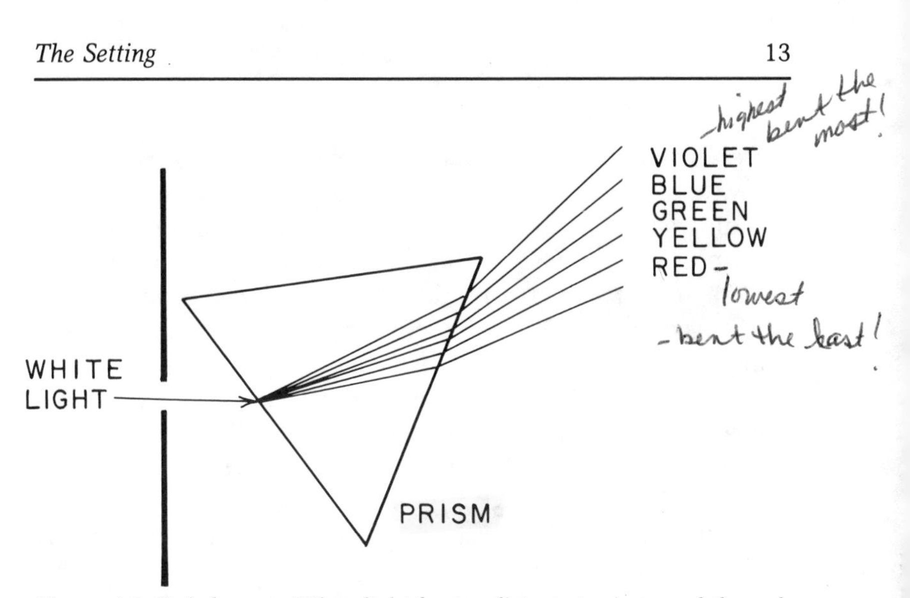

Figure 1-1. Rainbows: When light from a distant star is passed through a prism, the light is dispersed into the colors of the rainbow.

to lift one of the atom's electrons from one of its permitted energy levels to another. Using laboratory arcs as a means of calibration, astronomers were able to estimate the relative abundances of the elements making up the atmospheres of neighboring stars.

As bigger and better telescopes become available, astronomers attempted to extend their chemical analyses to more distant objects. It was here that the great discovery came. When astronomers looked at very distant objects they found that a shift occurred between the set of dark lines and the rainbow background. For example, a line that marred the blue part of a spectrum taken from the Sun would instead mar the green part of the spectrum for the light from a distant galaxy; a line that marred the yellow part of the Sun's spectra would mar the orange part of the distant galaxy's spectrum, and so forth. The spacing and relative intensity of lines remained the same. It looked as if someone had lifted the whole set of dark lines off the background rainbow, moved it toward the red end, and then replaced it. More startling was the finding that the more distant the object, the greater the shift toward red!

We can see why this is true if we first grasp a somewhat complicated concept. We will call it the train-whistle effect (physicists call it the Doppler shift). Those who have indulged in train-watching may remember that most engineers of express trains blow their whistles as they roar through local stations. Anyone standing on the platform experiences a strange sensation as the train passes. The pitch of the whistle suddenly drops! It drops for exactly the same reason that the lines in the spectra for distant galaxies shift. Since the whistle situation is a bit easier to comprehend, we will consider it first.

Sound travels through the air at a velocity of 740 miles per hour. If the train passes through the station at 74 miles per hour, then the frequency of the sound impulses on the listener's ears would be 10 percent higher as the train approached and 10 percent lower after it had passed by. This phenomenon is easily understood if we substitute for the train's whistle a beeper that gives off one beep each second. Were an observer to count the beeps from a train stopped down the tracks he would get 60 each minute. Were he to count the beeps from a train speeding toward him at 74 miles an hour, he would hear 66 each minute. Were he to count the beeps from a train speeding away from him at 74 miles per hour, he would hear only 54 beeps per minute. The ear counts the frequency of sound waves hitting our eardrum. When the source of the sound is approaching, the eardrum detects a higher frequency and sends to the brain a higher pitch.

If a source of light is receding, the "pitch" of its light is also lowered. However, as light travels at a staggering 670 million miles an hour, the frequency of light reaching us from a speeding train is not significantly changed. Thus, if we observe a shift toward red in the spectrum of the light reaching us from a distant galaxy corresponding to a 10 percent reduction in frequency, the galaxy must be speeding away from us at the amazing speed of 67 million miles an hour!

The Red Shift-Versus-Distance Relationship

As stated above, the more distant the galaxy, the larger will be the shift of its light toward red. Examples of actual spectra observed from a series of galaxies are shown in Figure 1-2. How do we know that the ones showing the greatest shift are those farthest away? Astronomers do this the same way you judge the

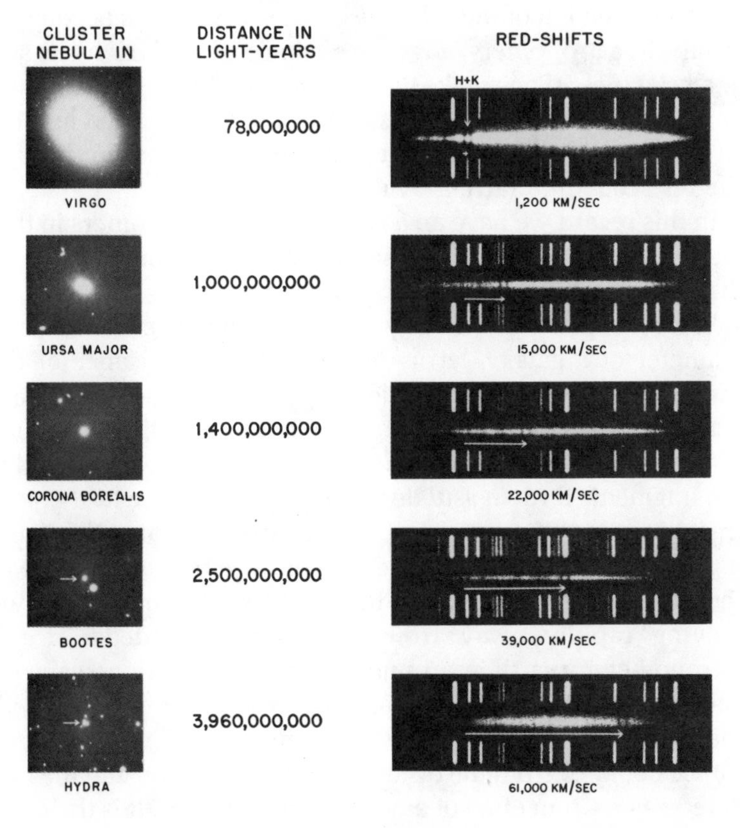

Figure 1-2. Galaxies and their light spectra: On the left are shown photos of five galaxies taken with the Hale Observatory telescope. As these objects are probably similar in size, Virgo must be located much closer to Earth than Hydra. Also shown, on the right, are light spectra from these galaxies. The white arrows show the displacement of an easily identified pair of dark lines from its position in a light spectrum for the Sun (or for a laboratory arc). The recession velocities corresponding to these arrow lengths are given. As can be seen, the more distant the object, the greater its recession velocity!

distance of an approaching car on a dark highway. As automobile headlights have similar luminosities and are roughly a standard distance apart, we judge the distance of an oncoming vehicle by a combination of the brightness and spacing of its headlights. Similarly, astronomers judge the distance of a galaxy by its size and brightness. They make the same assumption the driver does about headlights—that all galaxies are of similar size and luminosity. Thus, the smaller and weaker the galaxy as seen through the telescope, the farther away it is.

In this regard we have an advantage over astronomers in that we have seen many cars close up. The astronomer's sense of distance is calibrated in a far more complicated way.

Before exploring the way the astronomer measures the diameter of a galaxy, let us return to the significance of the distance-versus-red shift relationship. It tells us that the farther a galaxy is from us, the faster it is retreating. Although it may not be intuitively obvious, this is the relationship expected for the fragments thrown asunder by an explosion. The farther two fragments are from one another, the faster they must be moving apart. Since all the fragments were together at the time of the explosion, to have gotten far apart two fragments must be moving rapidly away from one another. Indeed, if the astronomer turns time around and runs the various galaxies backwards at the rates they are observed to be retreating, all these objects come together at the same time! The actual date can be obtained from the distance (from our galaxy) and the rate of recession (from ours) of any one of the galaxies. It is the same as the classic problem in which car A is speeding away from car B at the rate of 100 miles per hour. If the cars are 200 miles apart, then two hours ago they must have been at the same spot on the highway. In the case of galaxies, it is as if many cars left the same point at the same time. Each moved with a different velocity. Instead of traveling along the same highway, each follows its own highway out into 3D space.

Where is the center of the universe? A train analogy in Figure 1-3 shows why the red shift has nothing to say about this. Observers on night train A speeding along one track see a light

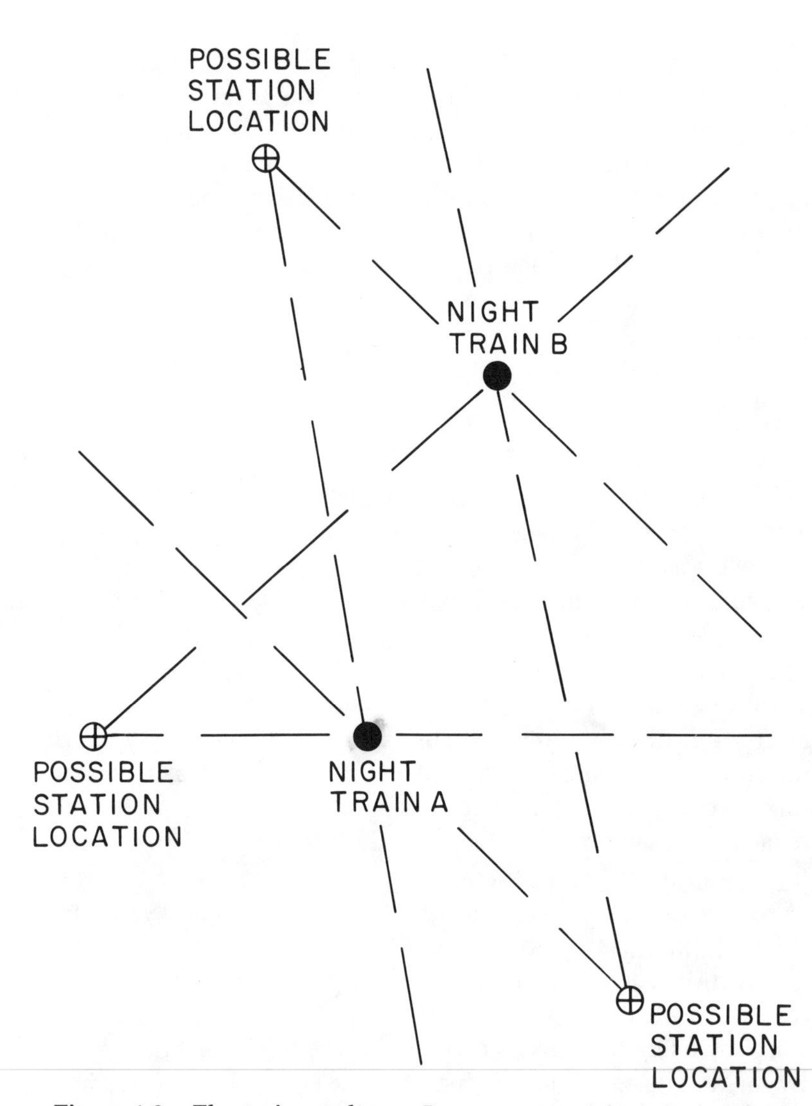

Figure 1-3. The train analogy: Passengers on night train A sight the beacon of night train B, which left the central station at the same time they did. From its brightness they determine the distance between the two trains. They also hear the whistle. From its pitch they determine the speed at which the two trains are moving apart. However, unless they have some other information (for example, the direction in which their train is going and the speed at which it moves along the track), there is no way for them to determine where the central station is located. Of the infinite number of possibilities, three are depicted here.

mounted on top of train B speeding along another track. They also hear train B's whistle. They know this train left the central station at the same time their train did. From the strength of its light they determine the distance of train B. From the pitch of its whistle they know that it is moving away from them and the exact speed of this recession. Knowing only this much the observers could not determine where the central station is located. Neither can astronomers locate the center of the universe.

Dating the Beginning

moving backwards

In order to determine the time at which the big bang occurred, astronomers had to measure the distance of a receding galaxy as well as its red shift. Distances are far more difficult to measure—so difficult that it is beyond our task here to try to grasp exactly how it is done. A few paragraphs will suffice to show the general principles.

As do all surveying schemes, measurements out into space start with a base line (see Figure 1-4). If a surveyor wants to measure the distance of an object that he cannot easily reach (like a rock out in a lake), he sets up a base line on shore and measures its length with a tape. He then observes the rock from both ends of the base line and notes the angle between the line of sight and the base line. Simple trigonometry allows him to calculate the distance to the rock.

As can be seen from Figure 1-5, the ranges of distance confronting the astronomer are incredible! The astronomer boldly starts by using the Earth's orbit about the Sun as his base line. By making observations of objects in the sky from the extremes of the orbit, the astronomer can use the triangulation method to measure the distance of "rocks" out in space. Even with this seemingly gigantic base line this proves to be a very tough task. The base line is 3×10^{13} centimeters long. Even the nearest star is 4×10^{18} centimeters away. Thus, it is like measuring the distance of a rock 10 kilometers (10^6 cm) off the coast using a

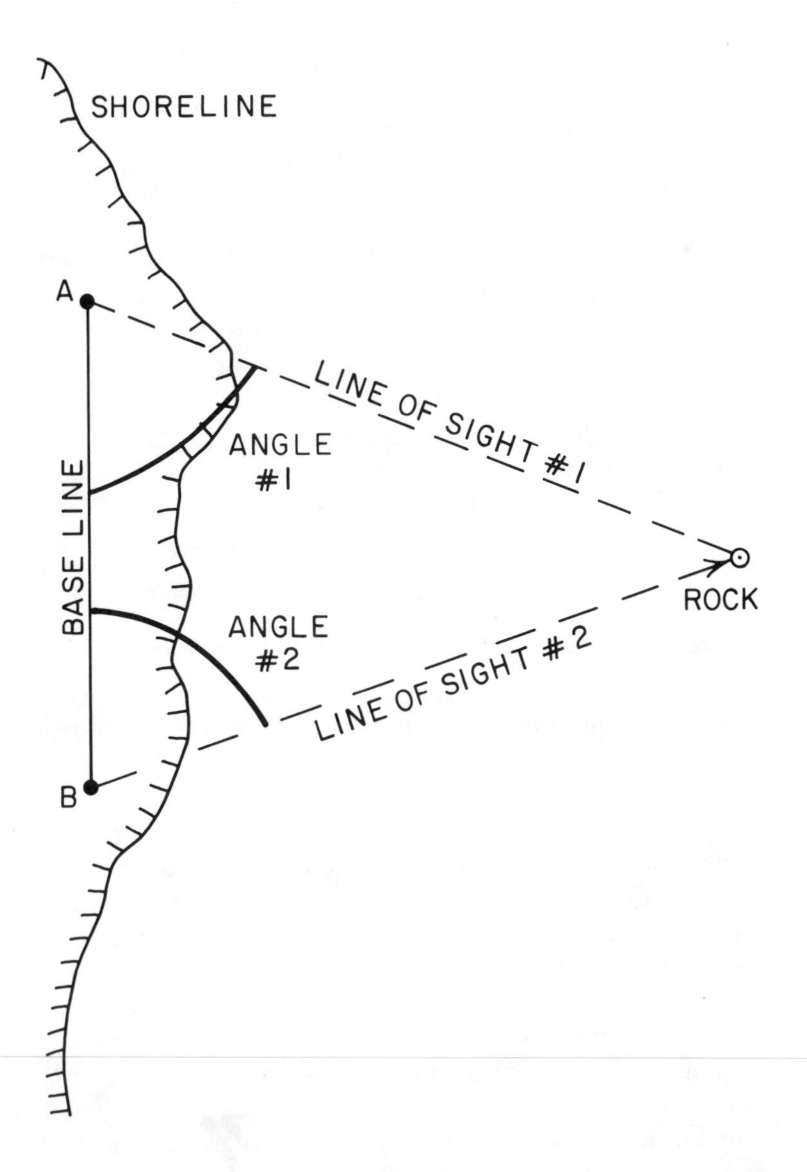

Figure 1-4. Surveying: Measuring the distance to an object without actually going to it is the goal of the surveyor. He observes the rock from the ends of a base line of known length and notes the angles between base line and the lines of sight. He can then compute the distance through trigonometry.

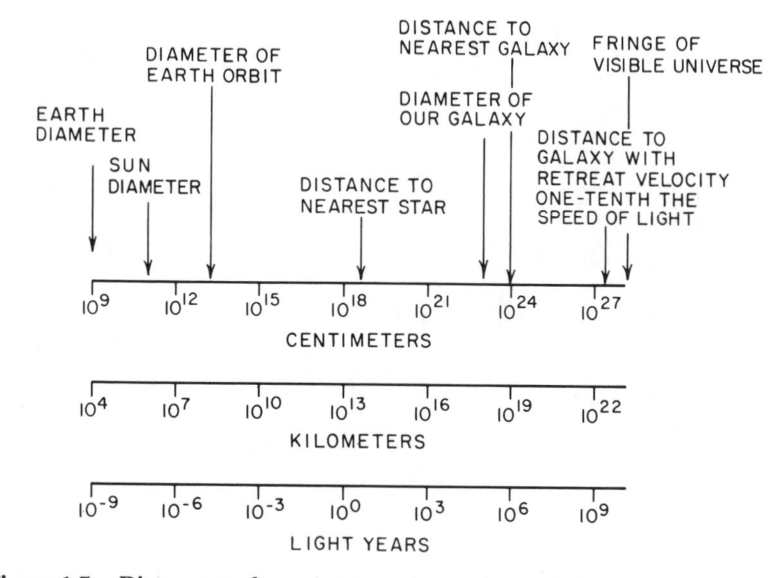

Figure 1-5: Distance scales: Astronomers must cope with distances ranging over 19 orders of magnitude! To measure these has been one of their major challenges. It takes 2.5 cm to make one inch. It takes 1.6 kilometers to make one mile. A light year is the distance light travels in one year.

base line only 1 cm long! Through the use of a very accurate technique called parallax, the astronomer can determine the distance to a few thousand of our nearest neighbor stars using the Earth's orbit as a base line. If he could go no farther in his distance observations, the astronomer would be limited to a very, very small portion of our own galaxy.

The astronomer's base line was greatly extended by showing that our own Sun is rushing through our galaxy at a rather large rate of speed. Through trigonometry he was able to show that the Sun moves 6×10^{13} centimeters through the galaxy each year. In this way he established an ever growing base line far longer than the Earth's orbit affords. It would be as if a surveyor were driving along a shore road in a golf cart periodically taking sightings on a distant island. From the cart's velocity and the elapsed time he could determine the length of his growing base line. In a related but more complicated way (called statistical parallax), astronomers have been able to measure the distances of stars out to about 3×10^{20} centimeters in our own galaxy.

Finding the distance to our nearest galactic neighbors posed a problem so formidable that the trigonometric approach had to be abandoned. Nature, however, provided an alternative approach which the astronomer hit upon and exploited. Some of the stars in our galaxy show regular pulsations in their luminosity. Hence, they are more like lighthouses than headlights. These stars show a whole range of blinking rates. The important thing, astronomers found, is that stars that blink at the same rate have the same luminosity. It's as if the Coast Guard decided to have all its lighthouses use "lightbulb" strengths related to their turning time. For example, all lighthouses with 100,000-watt bulbs would turn once each minute; those with 200,000-watt bulbs would turn once each two minutes, and so forth.

The astronomers jumped on this relationship and reasoned that blinking stars visible in nearby galaxies probably followed the same rules. In this way they calibrated their first "headlight" method. From the difference between the intensity of light received from one of these distant blinkers and the intensity of light received from one of its cousins in our own galaxy whose distance was determined by trigonometry, the distance to the nearby galaxies could be estimated. Knowing the distances of these nearby galaxies, the astronomer could determine their diameters. A "map" showing the Milky Way and its neighboring galaxies and clouds of gas and dust is shown in Figure 1-6.

Unfortunately, the galaxies that show significant red shifts are so far away that even our biggest telescopes cannot resolve individual stars. The whole galaxy presents an image nearly as small as that of a nearby star. Thus, the lighthouse method is not applicable. The last step out into space is taken by using the size of the galaxy itself. More often than not galaxies are found in clusters. Astronomers have carefully studied the sizes of the galaxies in nearby clusters. As with the sizes of people (and cars), they follow some simple rules. The assumption is made that the

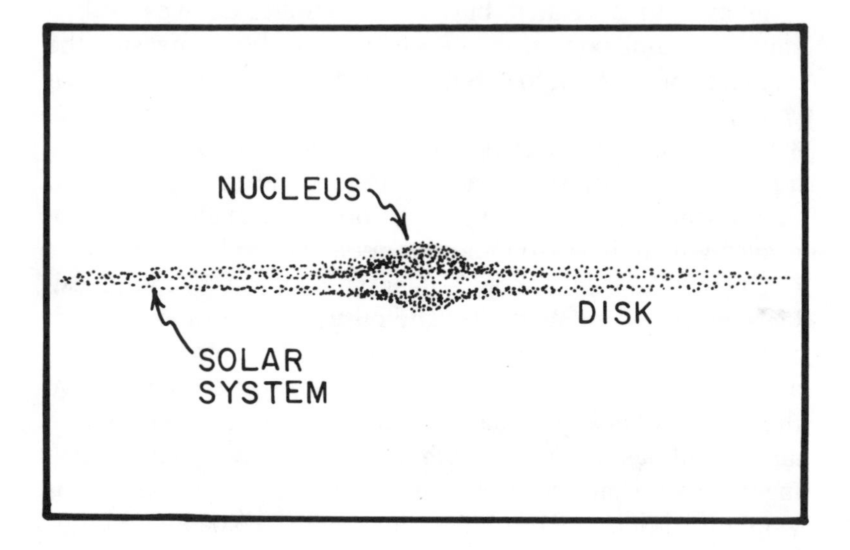

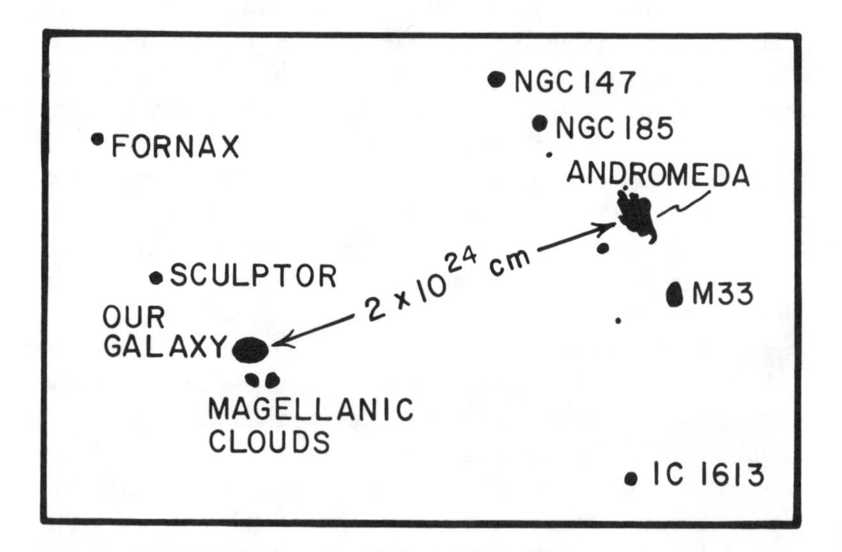

Figure 1-6. The Milky Way and its neighborhood: Drawings of what an alien astronomer looking toward our galaxy would see. Top, an edge-on drawing of our galaxy, the Milky Way. Bottom, a map of objects in the Milky Way's vicinity.

galaxies in very distant clusters have a similar spectrum of sizes and brightness as the ones in "nearby" clusters. Hence, like the automobile driver, the astronomer infers the distance of these clusters by the characteristics of their individual galaxies.

Having done this, astronomers can then make a graph with the distance of the galactic cluster on one axis and the velocity at which it is receding from us on the other axis. As shown in Figure 1-7, when the observations for various galactic clusters are plotted on such a graph, the points form a linear array. A factor of 10 increase in distance is matched by a factor of 10 increase in recession velocity. This relationship is to be expected if the matter in the universe is flying outward on the wings of an explosion that occurred 15 billion years ago. For example, a galaxy that is 4.6×10^{26} cm away from us is retreating at the rate of about 1×10^{9} cm/sec. If we turned things around and ran this galaxy back toward ours, it would take

$$\frac{4.6 \times 10^{26} \text{ cm}}{1 \times 10^{9} \text{ cm/sec}} \quad \text{or } 4.6 \times 10^{17} \text{ sec}$$

As there are 3.1×10^{7} seconds in a year, this comes out to be 15×10^{9} years! Because of uncertainties associated with the distance scale, there is an uncertainty of several billion years in this age.

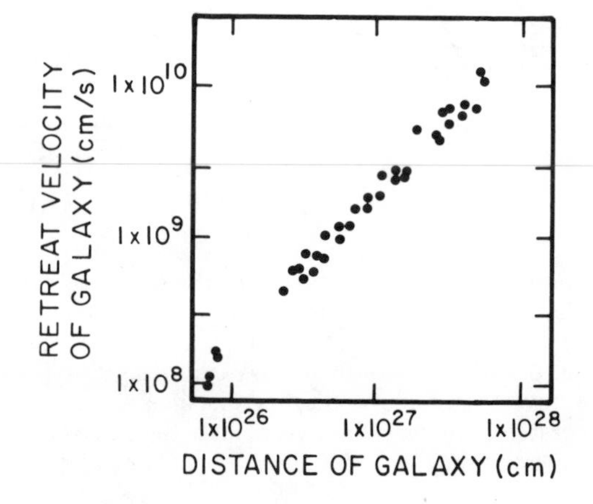

Figure 1-7. Relationship between galactic distance and galactic recession velocity: Each point represents a distant galaxy (or cluster of distant galaxies). Since the distances range over a factor of 100, a logarithmic rather than linear scale is used for this graph.

In Figure 1-8 the evolution of the distance-velocity relationship is depicted. Were we to have lived only 5 billion years after the big bang, the line depicting the velocity-distance trend would have been three times steeper than the one we obtain today. This is because the retreat velocity for any given galaxy remains nearly the same while the galaxy's distance from us increases.

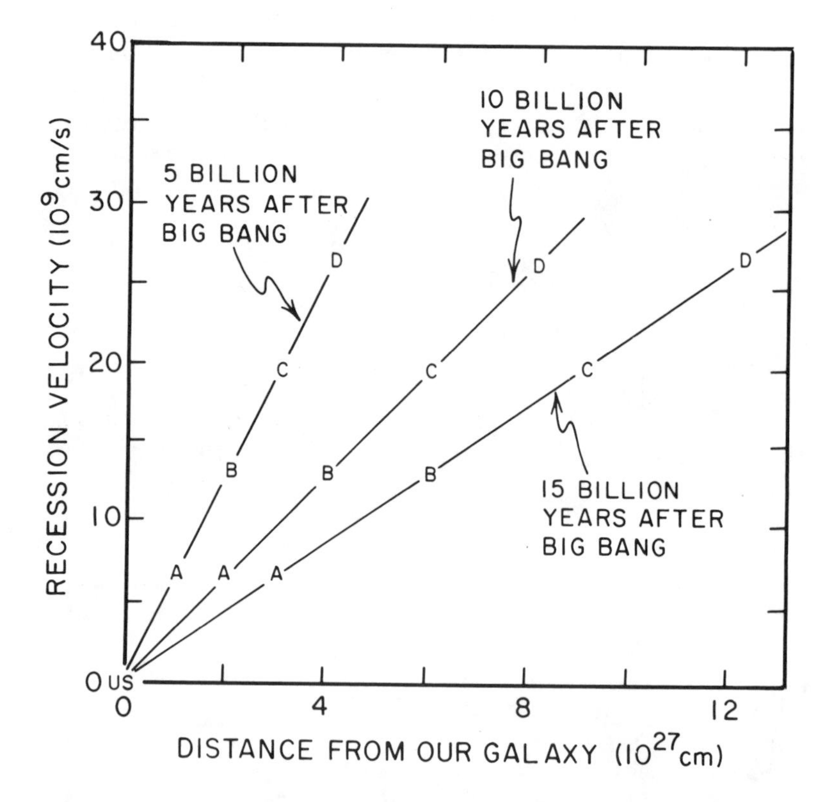

Figure 1-8. The evolution of the distance-versus-velocity relationship: Each of the four galaxies (A, B, C, and D) moves away from us at a different velocity. These velocities have remained nearly constant through time (that is, the velocity coordinate for each galaxy remains the same). However, as the universe becomes older, the distance separating these galaxies from us increases. Fifteen billion years after the explosion they are three times as far away as they were 5 billion years after the explosion.

Added Support for the Big-Bang Hypothesis

Physicists Wilson and Penzias of the Bell Laboratory in New Jersey won a Nobel Prize for showing that, although no visible light can be observed in the dark voids between stars and galaxies, there is a nonvisible glow to the universe. By viewing the universe with a detector sensitive to longwaved electromagnetic radiation in the 0.1- to 100-cm range (that is, microwaves), they were able to show that the universe has a background glow the same as that of an object at 2.76 degrees above absolute zero temperature (see Figure 1-9 for a description of the various temperature scales). After the Wilson and Penzias discovery, subsequent work demonstrated that the relative intensities of the various wavelengths of radiation in this range are consistent with this very cold universe temperature (see Figure 1-10). For comparison, the Earth gives off radiation characteristic of its surface temperature of about 288 degrees Kelvin. This radiation is centered in the infrared range. The Sun gives off radiation characteristic of its surface temperature of 5700 degrees. This radiation is centered in the visible range. A great flash of light appeared when the protons and electrons in the expanding universe cloud cooled to the point where they could combine into neutral atoms. At that time the universe was only about 100,000 years old and the gas had a temperature of about 4000 degrees Kelvin. The reason why this light, which was given off from a gas at 4000 degrees Kelvin, appears to have been given off by an object about 1500 times cooler (that is, one with a temperature of 2.76 degrees Kelvin) has to do with the expansion of the universe. While the computation of the magnitude of this "cooling" is too complex to be described here, to the physicist it is exactly as expected. Hence, the discovery of the afterglow of the big bang is taken by physicists as a strong confirmation of the exploding-universe hypothesis.

Aftermath of the Big-Bang

As we shall learn in the next chapter, after the big bang, universe matter consisted of only two elements, hydrogen and helium. About 100,000 years later, when the expanding matter had cooled to the point where the heretofore free electrons could become

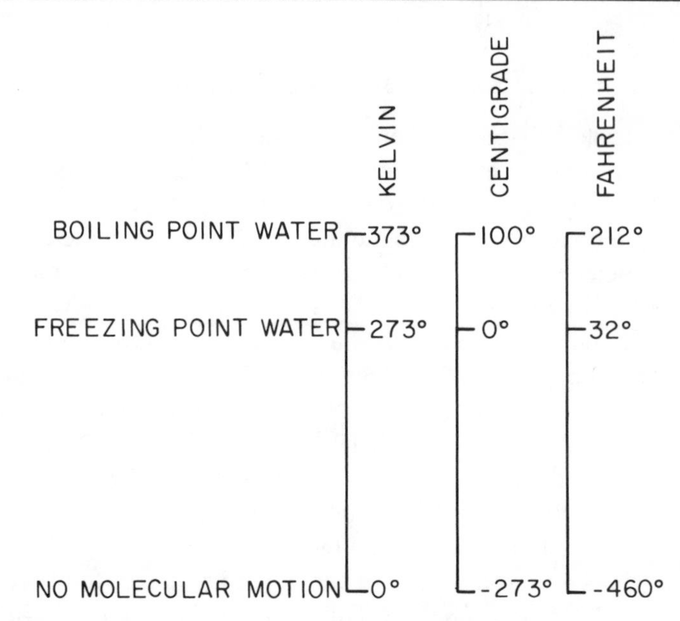

Figure 1-9. Temperature scales: Temperature is a measure of the intensity of molecular motion. When a substance is cooled to the point where its molecules come to rest, it is said to be at absolute zero temperature. When a substance has its molecular motions at equilibrium with those in an ice-water mixture, it has a temperature of 273 degrees on this absolute scale. When a substance has its molecular motions at equilibrium with those in a steam-water mixture, it has a temperature of 373 degrees. In everyday life we in the U.S. have long used the Fahrenheit scale. Recently there has been a shift to the Centigrade scale used by most other countries in the world. The relationship of these two scales of the weatherman to the absolute scale—or Kelvin scale—of the physicist is shown above.

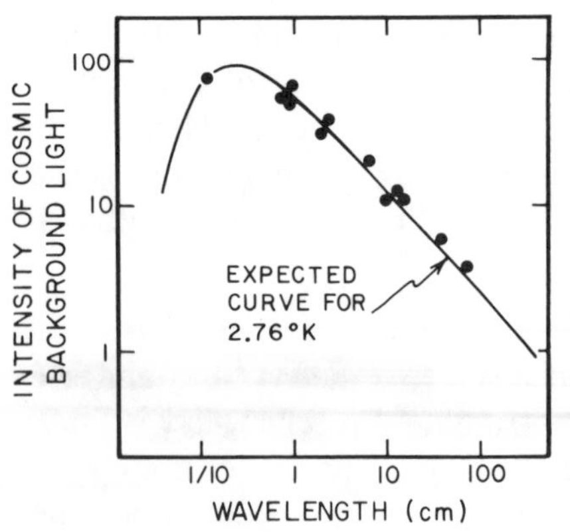

Figure 1-10. The cosmic microwave background: The universe glows with radiation in the microwave range. By measuring the intensity of this glow at many wavelengths, scientists have concluded that the spectrum corresponds to that of material with a temperature of 2.76 degrees above absolute zero.

entrapped in orbits around the positively charged nuclei, helium and hydrogen gas formed. This gas was lighted only by the afterglow of the big bang. At this point the universe was a dull place indeed. No galaxies, no stars, no planets, and no life were to be found. There were only molecules of gas in a rapidly expanding cloud.

Then, for reasons as yet not entirely understood, the cloud began to break up into a myriad of clusters. Once formed, these clusters remained as stable units bound by their mutual gravitation. Each of these clusters in turn evolved into one or more galaxies. Within these galaxies the gas further subdivided to form many billions of brightly burning stars. The universe was no longer dark!

While these early stars are by now either dead or lost among their younger counterparts, we can be quite sure that they had no Earth-like planets. The reason is that Earth-like planets cannot be formed from hydrogen or helium. Elements not present in the young universe are required. Thus, the next step will be to see where and how the remaining 90 elements formed.

Supplementary Readings

Frame of the Universe, *by Frank Durham and Robert D. Purrington, 1983, Columbia University Press.*

A history of thought regarding the structure and origin of the universe.

Galaxies and Quasars, *by William J. Kaufman, III, 1979, W.H. Freeman and Company.*

A discussion of the important units of matter making up the Universe.

The First Three Minutes, *by Steven Weinberg, 1977, Bantam Books, Inc.*

A book which puts the details of the early history of the universe within the grasp of the general reader.

The Big Bang, *by Joseph Silk, 1980, W.H. Freeman and Company.*

A book written for the nonspecialist which describes the achievements and puzzles in astronomy, cosmology and astrophysics.

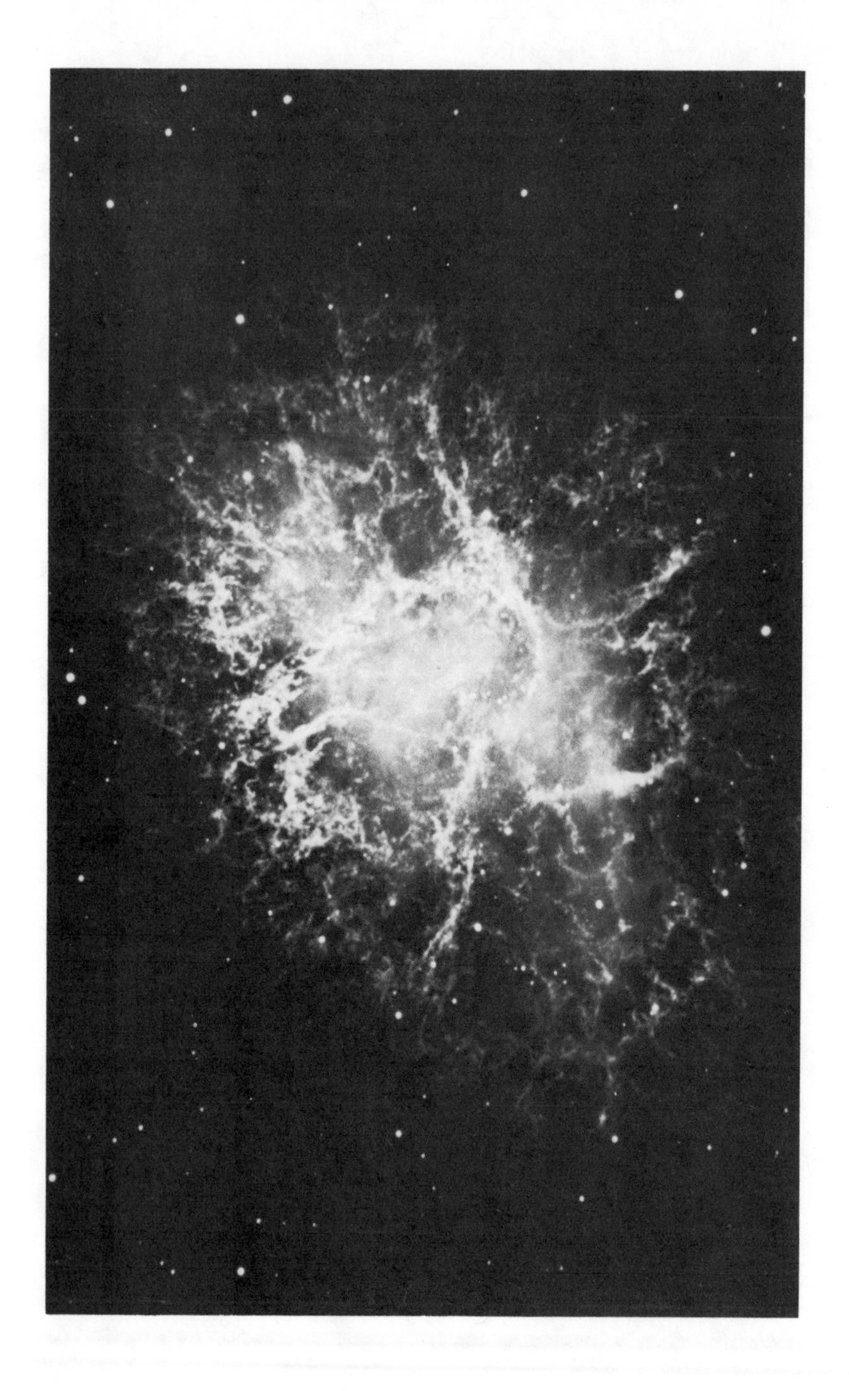

Crab Nebula

Chapter Two

The Raw Material:

Synthesis of Elements in Stars

During the explosive birth of our universe only two chemical elements were formed, hydrogen and helium. The other 90 elements were produced at a later time in the interiors of stars called "red giants." These massive stars rapidly burn their nuclear fuel and then explode, casting forth into the neighboring regions of the galaxy a mixture of the 90 missing elements. The frequency of these awesome events is about one per galaxy per century.

Evidence in support of this origin is imprinted in the relative abundances of the elements making up our solar system. For example, the high abundance of iron relative to its neighboring elements is consistent with the fact that iron is the ultimate product of the nuclear fires in large stars. That newly produced elements are present in supernova debris is demonstrated by the existence of the spectral lines of the element technetium in the light from the nebula formed by these explosions. Since all its isotopes are radioactive, technetium can be present only in matter fresh from a nuclear furnace.

Chinese astronomers observed a supernova in 1054. The debris cloud from this explosion is known as the Crab Nebula. Through the course of our galaxy's history, the formation and demise of about 100 million red giants has converted about 1 percent of the galaxy's hydrogen and helium into heavier elements. Contained in this 1 percent are the ingredients needed to build Earth-like planets.

Introduction

By cosmic standards our Earth and its fellow terrestrial planets are chemical mavericks. They consist primarily of four elements: iron, magnesium, silicon, and oxygen. By contrast, we look out on stars which are made up almost entirely of two elements, hydrogen and helium. For the universe as a whole, all elements other than hydrogen and helium are small potatoes; taken together they account for only about 1 percent of all matter.

Clearly, one of the prerequisites for habitability is that a planet have solid surface. Objects made primarily of hydrogen and helium gas offer no such base. Hence, high on our agenda must be an understanding of how elements heavier than hydrogen and helium were formed and how these elements were separated from the bulk gas and forged into rocky planets. In this chapter we deal with the first of these problems; in the next chapter we deal with the second.

The Chemical Composition of the Sun

All stars form from the gravitational collapse of clouds of gas. Since the lion's share of the hydrogen and helium in the collapsing clouds ends up in the star itself, the star's chemical composition must be representative of the parent cloud. Thus, if we could somehow determine the chemical composition of the Sun, we could define the composition of the galactic matter from which the Sun formed. As there is no reason to believe that our galaxy is atypical, such an analysis would give us a rough idea of the composition of average universe matter.

Our information about the composition of stars comes from the dark lines in "rainbows" generated by passing their light through a diffraction grating (a device which, like a prism, is capable of separating the various colors into a spectrum). As was discussed in the previous chapter, the light emitted from the white-hot Sun is partially absorbed by the gases in its atmosphere. This absorption mars the otherwise continuous gradation in color. Each line that mars these rainbows is characteristic of a single element. The extent to which the frequency of light corresponding to each line is deleted from the rainbow is a measure of the abundance of that particular element in the Sun's atmosphere. Fortunately for small stars like our Sun, the atmosphere is thought to have a composition nearly identical to that of the interior.

By passing the light from electrical arcs through gas mixtures of known composition in the laboratory, physicists have been able to learn which lines correspond to which elements and how the relative degree of darkening at the positions of those lines relates to the relative abundances of elements in the gas around the arc. This knowledge was applied to the reading of solar rainbows and allowed astronomers to obtain estimates of the relative abundance of most of the elements in the Sun's atmosphere. By "relative abundance" we mean the ratio of the number of atoms of a given element to the number of atoms of a reference element. By convention, astronomers use silicon as the reference element. The relative abundance of an element is stated as the number of atoms of that element for each 1 million atoms of silicon. These abundances are plotted versus element number on the graph in Figure 2-1. This graph has a power-of-ten scale. For example, helium atoms with a relative abundance lying between 10^9 and 10^{11} on the scale are 10 billion times more abundant than bismuth atoms with a relative abundance lying between 10^{-1} and 10^0.

Other than the dominance of the abundances of hydrogen and helium over those of the other 90 elements, the most obvious feature of the graph is the general decline in abundance with increasing element number. Superimposed on this decline are two anomalies of particular note. One is that the abundance of the element iron is 1000 times higher than would be expected for a smooth decline. The other is that the elements lithium, beryllium, and boron have abundances many orders of magnitude lower than would be expected if the decline were smooth. In addition to these two very prominent anomalies, there are numerous small ones which give the abundance curve a sawtoothed appearance. For example, elements with an odd number of protons are generally less abundant than the neighboring element with an even number of protons. As we shall see,

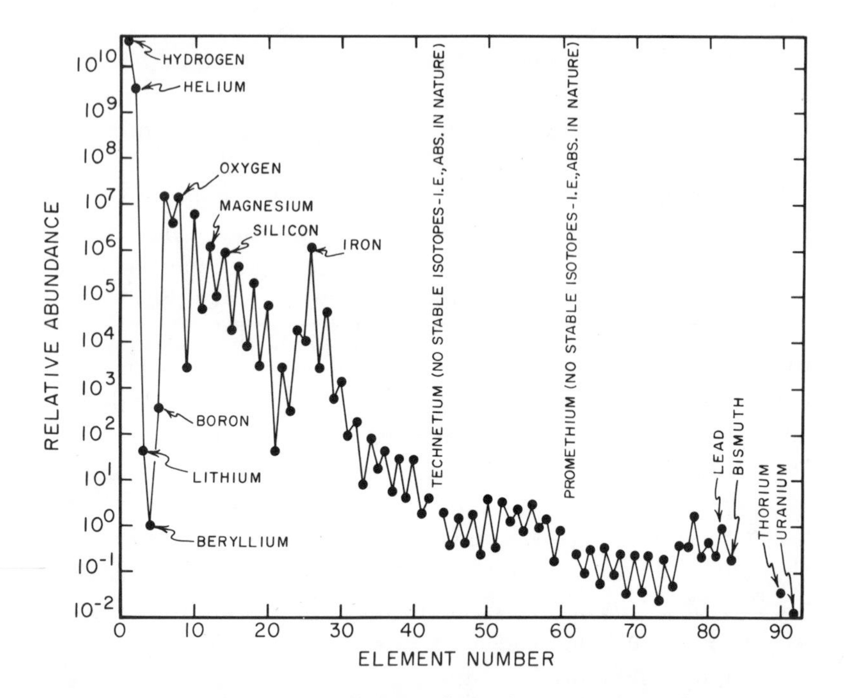

Figure 2-1. Relative abundances of the elements in our Sun: As the abundances range over 13 orders of magnitude, they must be displayed on a power-of-10 scale. The abundance of each element is expressed as the number of atoms per million (i.e., 10^6) atoms of the element silicon. The gaps in the sequence represent elements that have only radioactive isotopes and are, therefore, absent in the Sun. While most of the abundances are based on spectral data, use is made also of chemical measurements on a special class of meteorites called carbonaceous chondrites.

these characteristics of the abundance curve provide hot clues regarding the mode of origin of the elements heavier than hydrogen and helium.

Physicists conjecture that at the instant of the big bang, all matter must have been in a very compact blob. The pressures and temperatures in this primordial blob were so high that stable combinations of neutrons and protons could not exist. Within seconds after the explosion, however, such combinations could and did form. At one time it was thought that the blend of elements we see in our Sun might have been generated entirely during the first hour of universe history. But subsequent work has shown that the only elements produced in significant amounts during this very early phase of universe evolution were hydrogen and helium. The others were produced billions of years later inside giant stars.

The hydrogen and helium gas produced during the big bang eventually agglomerated into megaclouds. These megaclouds organized into the spiral and elliptical shapes we see in distant galaxies. Some of the gas in these newly formed galaxies in turn broke into far smaller subclouds which collapsed under their mutual gravitation into stars. So (through their telescopes) astronomers see a host of galaxies, each defined by billions of twinkling stars. Through careful observation astronomers have been able to show that the process of star formation still goes on. They see new stars forming and old ones dying. By observing stars of all sizes and in all stages of their evolution, astronomers have been able to map out the history of these objects. Interwoven with this evolution is the conversion of hydrogen and helium to heavier elements. It is here, rather than in the big bang, that we must look for the production of the iron, magnesium, silicon, and oxygen of which Earth is comprised.

Again one might ask, how do scientists know that the elements heavier than helium were born in the centers of stars? As we shall see, a rather impressive case can be made. Most juries would believe it. Like the big-bang theory, the stellar synthesis theory of element origin gets a ranking of 9 out of a possible 10.

Descriptive Nuclear Physics

To appreciate the case to be put forth in defense of the stellar synthesis hypothesis, you must learn a bit about nucleonics. The term *nucleonics* is used rather than *nuclear physics* because we need to consider only some of the very simple facts about the nuclei of atoms. Just as the user of a pay telephone need not understand the intricacies of the switching system that routes calls to the desired places, we need not delve into the physics of nuclear architecture to appreciate what goes on inside stars.

Each atom has a compact nucleus made of neutrons and protons. This nucleus carries nearly all of the atom's mass and is incredibly small, only about 10^{-13} cm in diameter. A fluff of electrons flying in complicated orbits around the central nucleus gives the atom its bulk but adds almost nothing to its mass. The diameter of this electron cloud is about 10^{-8}cm (that is, the atom is 100,000 times bigger than its nucleus). The electrons, which have negative electrical charges, are held captive in orbit by the electrical attraction of the positively charged protons in the nucleus.

The reactions of interest to the chemist involve the sharing of electrons by two or more atoms. Through sharing, atoms are bound together into chemical compounds. During a chemical reaction, only the character of electron orbits changes; the nucleus remains intact. In contrast, physicists are interested in reactions by which the nuclei of the atom are altered. Like the ancient alchemist, the physicist seeks ways to turn lead into gold. It is here that our interest lies. How can the magnesium, iron, oxygen, and silicon atoms that make up the earth be manufactured from hydrogen and helium?

Most chemical reactions require heat. To make Jello we need hot water. To start a fire we need a spark. To make ceramics or steel we need white-hot ovens. Most nuclear reactions also require heat. A major difference exists, however, in the intensity of the required fire. Atoms can be enticed to undergo chemical reactions at temperatures in the range of hundreds of degrees centigrade to thousands of degrees centigrade. Such temperatures can easily be achieved. Early humans created them by striking pieces of flint together. You perhaps have done it by using a magnifying glass to focus the Sun's rays on a piece of paper. In stoves it is done by passing electricity through metal elements. To ignite the nuclear fires of interest to us there must be temperatures of 50 million degrees or more. Generating such temperatures is no simple task. Only by accelerating charged particles in mighty machines or by setting off nuclear explosions can physicists create these high temperatures. This is why the alchemists dedicated to making gold from less valuable elements failed. They had no means by which to start a nuclear fire!

The only places in the universe where there are natural furnaces with the temperatures required for nuclear fires are at the centers of stars. Every star must have such a fire at its core, otherwise the star would not shine. Thus, if alchemy is occurring anywhere in the universe, it must be in the cores of stars.

To understand which nuclides might be manufactured in stars we must be aware that only certain combinations of neutrons and protons form stable units. As shown in Figure 2-2, out of all the possible combinations relatively few are in this stable category. The rest are radioactive and, given enough time, will spontaneously transform into one of the stable combinations. The pathways for these transformations are shown in Figure 2-3. We will focus our attention on those nuclides which, left to themselves, will remain forever.

As can be seen in Figure 2-2, the stable nuclide with the most neutrons and protons (that is, the biggest and heaviest nuclide) is ^{209}Bi. All nuclei with more than 209 particles are radioactive. It can also be seen in Figure 2-2 that the stable nuclides form a band running from ^{1}H at one end to ^{209}Bi at the other. The course of this band represents the most favorable ratio of neutrons to protons. This ratio is near unity for the low proton number elements and rises, with proton number reaching 1.5 for bismuth. One might say that the amount of neutron "glue" required to hold the protons together gets larger as the nucleus gets larger. Exactly why neutrons are able to overcome the powerful electrical repulsion between protons, permitting neutrons and protons to coexist in an ultradense package, is understandable only to the nuclear physicist.

All stable nuclides are found in nature. Thus, all must somehow be produced from hydrogen and helium at the centers of stars. As we shall see, this buildup from small to large occurs in many steps. To produce a carbon atom requires only two steps; to produce an iron atom requires a few more steps; to produce a bismuth atom requires many more steps. It is because of this stepwise buildup that the "light" elements were produced in greater abundance than the "heavy" elements.

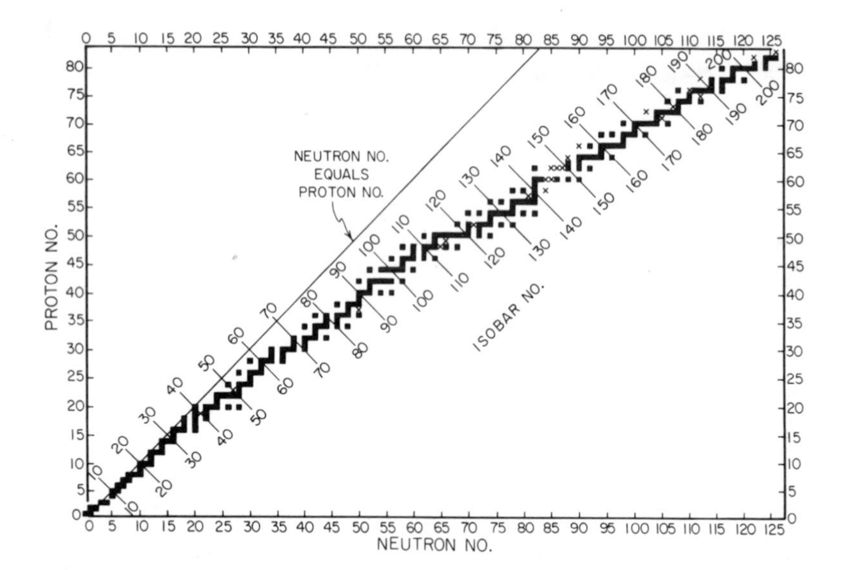

Figure 2-2. Stable nuclide configurations: The squares represent stable combinations of neutrons and protons. The x's represent radioactive nuclides whose half-lives are so long that they survive billions of years after their formation in stars. All the remaining combinations are radioactive with half-lives sufficiently short that they are no longer present in the solar system. Nuclides lying along the same horizontal line (i.e., those with the same proton number) are referred to as isotopes. Those falling along the same vertical line (i.e., those with the same number of neutrons) are referred to as isotones. Those falling along the same diagonal line (i.e., those with the same number of nuclear particles) are called isobars. The diagram terminates with the heaviest stable nuclide (^{209}Bi).

Element Production During the Big Bang

Let us see what these steps are. As has already been stated, the universe began as a blob of matter. In the fireball of the big bang this matter was largely in the form of neutrons. Once released from their dense confinement, neutrons are able to undergo spontaneous radioactive decay to protons. The decay process of a neutron involves breakup into one proton and one electron. The half-life for this decay is 12 minutes. Hence, 12 minutes after the universe formed, matter consisted of about half neutrons and half protons. In the still moderately dense

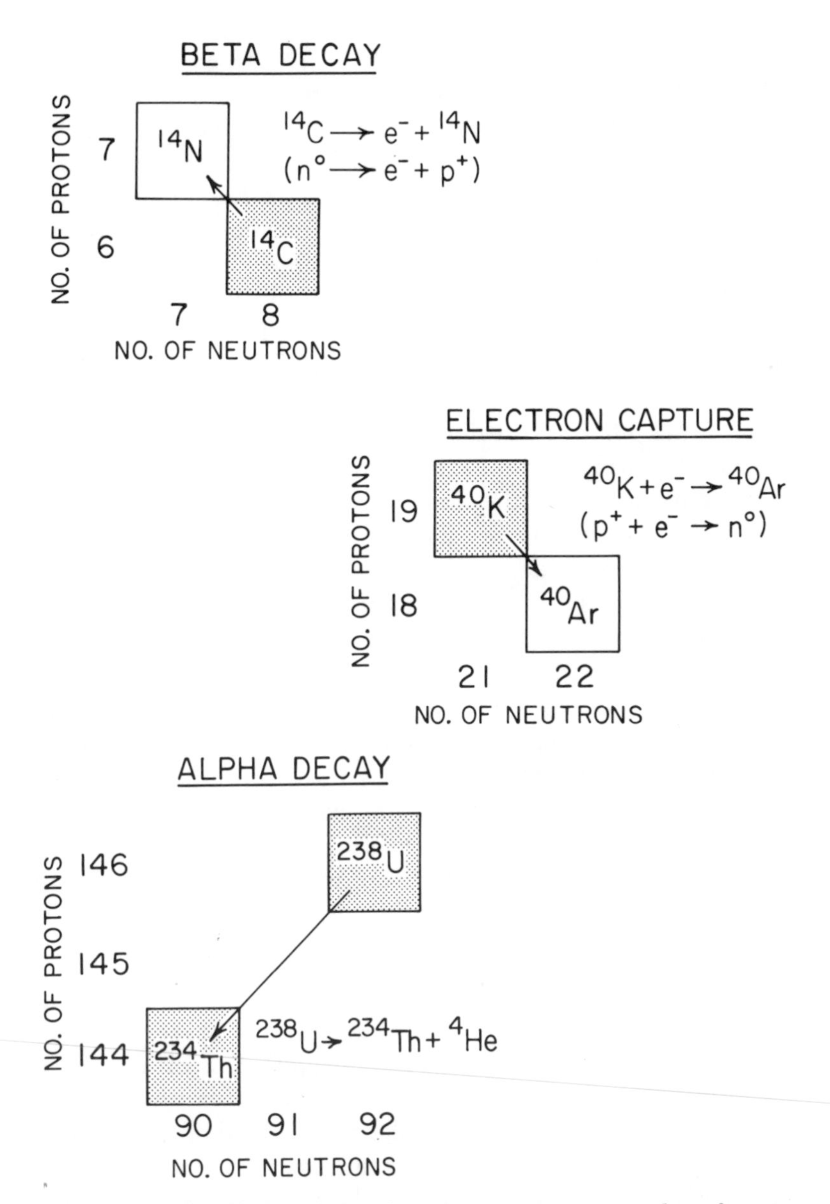

Figure 2-3. Examples of the three most common modes of spontaneous radioactive decay: Two of these, beta decay and electron capture, are isobaric—i.e., the number of nucleons remains the same. The third, alpha decay, involves the ejection from the nucleus of four particles in the form of a ^4He nucleus.

mass of rapidly expanding material frequent collisions occurred which led to the formation of helium gas. While these collisions might also have produced elements heavier than helium, to do so would have required that three nuclei simultaneously merge. The reason is that there is no stable nuclide with five nuclear particles (see Figure 2-4). For example, two protons would have to have collided at the same time with a ^{4}He nucleus to produce ^{6}Li. As is the case on a pool table, in the expanding gas cloud "three-ball" collisions were far less frequent than "two-ball" collisions, so infrequent, in fact, that the number of nuclei formed that were heavier than ^{4}He was insignificant. Thus, for all practical purposes, at the end of Day One universe matter consisted entirely of the elements hydrogen and helium. Further synthesis awaited the formation of galaxies and the formation of stars within these galaxies.

Physicists have made models of the collisions that would have occurred during the first day of universe history. They find that about 24 percent of the matter in the universe should have been converted to ^{4}He (the other 76 percent remained as bare protons formed by the decay of the original neutrons).* The fact that this is about the fraction of helium seen in young stars from throughout the universe is taken by these physicists as added evidence in support of the big-bang hypothesis. If there were considerably more or considerably less helium relative to hydrogen then all would not be well with the big-bang hypothesis!

Element Formation in Stars

Stars are hot inside for the same reason that brake shoes on a stopping car are hot. When a moving vehicle is brought to a stop, the energy associated with its motion is converted to heat in its brake linings. During the collapse of a cloud of gas, gravitational energy is also converted to heat. The amount of heat produced is so vast that the core of the protostar becomes hot

* While there are 60 ^{4}He atoms for every 1000 ^{1}H atoms, because the helium atoms are four times as massive, they account for 24 percent of the universe mass.

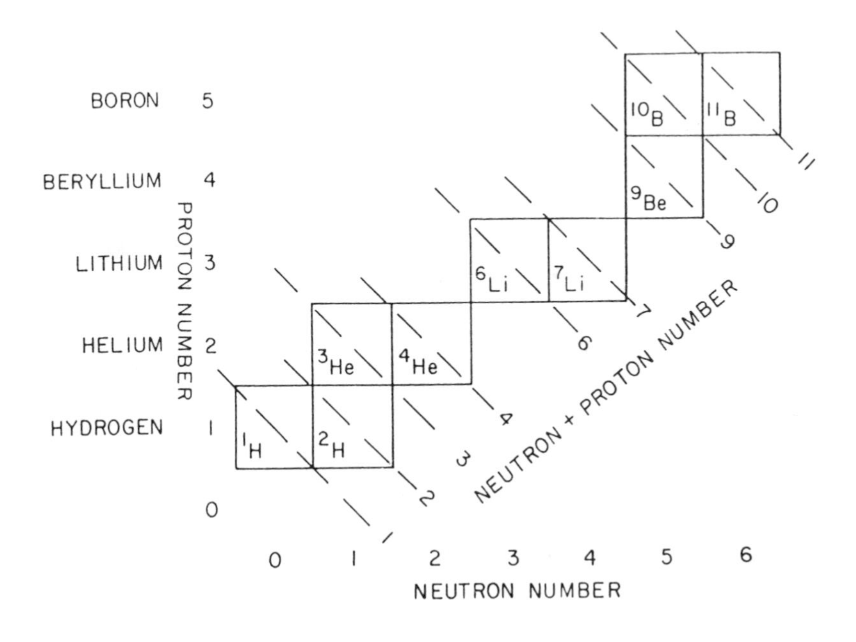

Figure 2-4. Stable nuclides with a particle number in the 1-to-11 range: Note that no stable nuclide exists with neutron-plus-proton number totaling 5 or 8. It is these two gaps in the chain that prevented element formation during the big bang from continuing beyond helium.

enough to ignite a nuclear fire.

For the nuclei in a star to react, they must touch. To touch, they must fly at one another at such high velocities that they overcome the electrical repulsion exerted by one proton on another. It is much like trying to throw a Ping-Pong ball into a fan. A very high velocity is required to prevent the ball from being blown back in your face.

The hotter atoms are, the faster they move. As stated in Chapter One, temperature is a scale for molecular motion. Touching a hot stove causes the molecules in the skin of your finger to move so fast that the chemical bonds holding them in place are rent; we call this molecular damage a burn. For two protons to collide requires velocities equivalent to a temperature of about 60 million degrees centigrade. Through a somewhat complicated series of collisions, four protons can combine to produce a helium nucleus (and two electrons). The helium nucleus contains two of the original protons and two neutrons. These neutrons come into being through the mergers of protons with electrons (for each proton in a star there must be one electron).

As first recognized by Einstein, for a nuclear reaction to lead to a release of heat there must be a reduction in the mass of the constituent atoms. This lost mass reappears as heat. Indeed, the weight of a helium atom is just a little less than that of four hydrogen atoms (see Table 2-1). Thus, for each helium atom manufactured in a star, a little mass is converted to heat. As the proponents of fusion power are quick to point out, the amount of heat obtained in this way is phenomenal. So phenomenal, in fact, that once a protostar's nuclear fire is ignited, its collapse is stemmed by the pressure created by the escaping heat. The star stabilizes in size and burns smoothly for a very long time. For example, our Sun has burned for 4.6 billion years and won't run short of hydrogen fuel for several more billion years.

Most of the stars we see are emitting light created by the heat from a hydrogen-burning nuclear furnace. Thus, one might say that stars are continuing the job begun during the first day of universe history; they are slowly converting the remaining hydrogen in the universe to helium.

As hydrogen is a universe resource good for many tens of billions of years, there is no need to call on any fuel other than hydrogen. However, if hydrogen alone burns and helium alone

Table 2-1. Conversion of mass (m) to energy (E):

Einstein is famous for his equation: $E = mc^2$, where c is the velocity of light. If four hydrogen atoms are converted to one helium atom, the mass loss is as follows:

Mass of 4 hydrogen atoms	6.696×10^{-24} gm
Mass of 1 helium atom	-6.648×10^{-24} gm
Mass loss	0.048×10^{-24} gm

Using Einstein's equation, this mass loss generates about 1×10^{-12} calories of energy. If one gram of hydrogen is converted to helium, then about 1.5×10^{11} calories of energy are produced. With this amount of energy, about 2 million liters of water could be heated from room temperature to the boiling point.

is generated, then there is no way to explain the presence of the other 90 elements found in nature.

Other fuels are burned. In large stars the supply of hydrogen is exhausted on a far shorter time scale than in our Sun (by stellar standards our Sun is a small-fry). So-called red giants run through their hydrogen supply in something like a million years. When the core of a red giant becomes depleted in hydrogen, the nuclear fire dims and the star loses its ability to hold back the pull of gravity. It once again begins to collapse. The energy released by the renewed collapse causes the core temperature to rise. Eventually the ignition temperature of helium is reached. Since helium nuclei have two protons, the force of electrical repulsion between them is four times the force of repulsion between two hydrogen nuclei. Thus, it takes a much higher temperature to ignite helium than is needed to ignite hydrogen.

Once this new temperature threshold has been reached, helium nuclei begin to combine to form carbon nuclei (three ^4He nuclei merge to form one ^{12}C nucleus). The mass of the carbon atom is less than that of the three helium atoms from which it was formed. This lost mass appears as heat. The heat from the rekindled nuclear fire stems the star's collapse and its size once again stabilizes. Oxygen atoms form as well. Four ^4He nuclei combine to make one ^{16}O.

The first red-giant star to form in the universe manufactured carbon, an element absent from the universe up to that time! While not a major component of the Earth, carbon is the primary constituent of living organisms. Also, it plays a big role in controlling climate.

In the largest stars this cycle of fuel depletion, renewed collapse, core temperature rise, and ignition of a less flammable nuclear fuel is repeated several times (see Figure 2-5). Carbon nuclei merge to form magnesium and so forth. Each merger leads to a small loss of mass and to the corresponding production of heat. The buildup cannot go beyond the element iron, however, because beyond iron heat must be *added* to the nuclei if they are to merge. The mass of nuclei heavier than iron proves to be slightly *larger* than the mass of the nuclei that are merged to form them. Thus, the nuclear furnaces of stars can produce only elements ranging from helium through iron. It should be noted that included in this range are the elements oxygen, magnesium, and silicon. Thus, if there were a means by which these elements could escape from the centers of the large stars, we would have the four major ingredients needed to build Earth-like planets.

Before discussing the escape mechanism, let us briefly consider the fate of more ordinary stars like our Sun. When the core of our Sun runs out of hydrogen, several billion years from now, it will resume its collapse. However, our Sun is just barely massive enough to generate the temperature necessary to start a helium fire. Then, after it has burned a bit of its core helium, it will collapse into a very dense object which will cool slowly until it gives off only a dull glow. A star to which this has already happened is referred to as a "white dwarf."

Element Synthesis by Neutron Capture

In contrast to small stars, which experience quiet deaths, big stars experience violent deaths. When the last of their core's nuclear fuel is consumed, the ensuing collapse is catastrophic. Rather than rekindling a controlled fire, the collapse becomes an implosion which tears the star asunder. The products of the earlier fires are cast into the surroundings. Astronomers call these implosions supernovae.

Nuclear reactions occurring during supernovae create elements heavier than iron. To understand these reactions we must consider the one nuclear reaction that can occur at "room temperature." It is called neutron capture. Because the neutron has no charge, it is not repelled by any nucleus it happens to

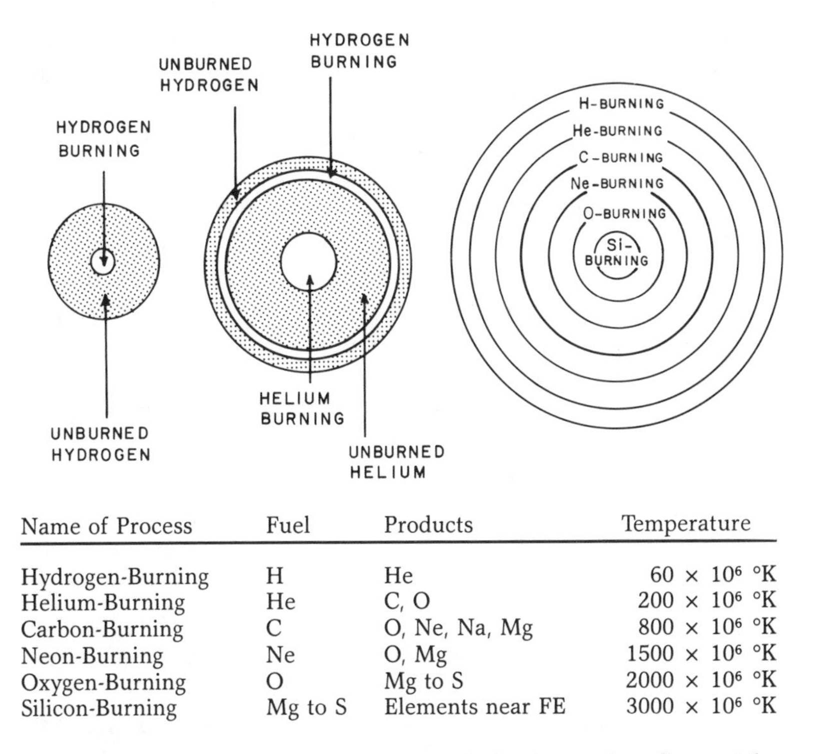

Name of Process	Fuel	Products	Temperature
Hydrogen-Burning	H	He	60×10^6 °K
Helium-Burning	He	C, O	200×10^6 °K
Carbon-Burning	C	O, Ne, Na, Mg	800×10^6 °K
Neon-Burning	Ne	O, Mg	1500×10^6 °K
Oxygen-Burning	O	Mg to S	2000×10^6 °K
Silicon-Burning	Mg to S	Elements near FE	3000×10^6 °K

Figure 2-5. Three stars with progressively hotter nuclear fires: Like our Sun, the star at the left burns hydrogen to form helium in its core; this core is surrounded by unburned fuel. The middle star is burning helium to form carbon and oxygen in its core. This core is surrounded by a layer of unburned helium. Outside of this is a layer in which hydrogen burns to produce helium. Finally there is an outer layer of unburned hydrogen. The star on the right has a multilayered fire. The successive nuclear fires are separated by layers in which no reaction occurs. These layers contain the same fuel as is being consumed in the underlying fire. These layers are depleted in the ingredient being consumed in the overlying fire. The approximate temperatures required to ignite the successive fuels are also given.

encounter; it can freely enter any nucleus regardless of how slowly it is moving. This ability of the neutron to react with nuclides under "room temperature" conditions lies at the heart of the principle of the atomic bomb and of the nuclear power reactor.

During the explosion that marks the death of a red giant, a host of nuclear reactions occur which create free neutrons. Thus, during these events the neutral particles which dominated the universe at its very beginning briefly reappear. The importance of the reappearance of neutrons is that they provide a means of building elements heavier than iron. In the close-packed conditions that exist inside stars, the neutrons created during nuclear reactions encounter a nucleus long before they get around to undergoing spontaneous decay to a proton plus an electron. Many of these encounters will be with iron nuclei. The iron nucleus absorbs the neutron and thereby becomes heavier. In the supernova explosion these neutron hits will be like bullets from a machine gun. No sooner is an iron atom hit by one neutron than it is hit by another. The iron atom gets heavier and heavier until finally it cannot absorb any more neutrons. Additional hits pass right through. This very brief pause in growth ends when the iron nuclide emits an electron (that is, when it undergoes radioactive decay). In so doing, one of its neutrons becomes a proton and the iron atom becomes a cobalt atom. The cobalt atom will in turn absorb neutrons one after another until it too becomes saturated. It then emits an electron and in so doing becomes a nickel atom. Two steps have been taken along the road from iron to uranium.

This sequence is repeated over and over again, driving matter along the neutron saturation route (see Figure 2-6). Because of the rapidity of the impacts, radioactivity proves to be no barrier to this buildup. The buildup zooms past bismuth and even past uranium and thorium, stopping only when the nuclei get so big that the neutron impacts are able to cause the nuclide to fission. The fragments produced by a fission event are caught up in the bombardment and begin once again to move along the saturation route.

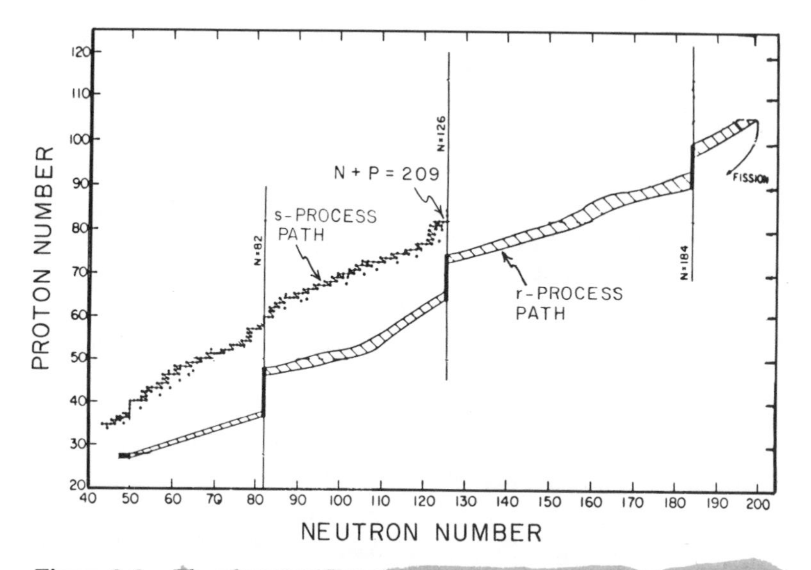

Figure 2-6. The elements heavier than iron were built by neutron irradiation: Two quite different processes contributed to this production. One, the s-process (i.e., slow process), occurs concurrently with the production of iron in the stellar core. As in a nuclear reactor, the reaction proceeds in a controlled way. Neutron hits are spaced out in such a way that the nuclides have time to achieve stability through beta decay. Thus, the buildup path follows the belt of stability shown in Figure 2-2. For the same reason it terminates at ^{209}Bi, the heaviest stable nuclide.

The r-process (i.e., the rapid process) occurs during the supernova explosion. Thus, it is akin to an atomic bomb. No sooner has a nuclide absorbed one neutron than it is hit by another. No time exists between hits for radiodecay. Rather, radiodecay occurs only when the nuclide becomes so neutron-rich that it cannot absorb any more. This leads to a buildup path displaced from the stability belt as shown. It also allows the buildup to proceed beyond particle number 209. Instead, the buildup goes just beyond particle number 300. At this point the colliding neutrons cause the nuclides to fission. The jogs in the r-process pathway occur at the so-called magic neutron numbers, 82, 126, and 184. They are "magic" in the sense they give the nuclide unusual stability.

Since we are dealing with an explosion, this bombardment is as brief as it is intense. As suddenly as they appear, the neutrons disappear. The host of neutron-saturated nuclei are now free to undergo radiodecay. These neutron-rich isotopes emit one electron after another until they have achieved the stable neutron-to-proton ratio (see Figure 2-7). Those nuclides heavier than bismuth emit alpha particles (helium nuclei) as well as electrons, moving toward stability as lead isotopes. While for most nuclides this adjustment process is quickly completed, in a few cases where nuclides with long half-lives for radiodecay lie in the sequence leading to stability, the adjustment process still goes on today. As we shall see, the radioactivity of these remaining long-lived isotopes plays a very important role in the evolution of individual planets.

While producing nuclides ranging in mass from 56 (iron) up to 238 (uranium), the so-called r-process does not produce all the stable nuclides in this mass range. As can be seen in Figure 2-2, there are generally two stable nuclides on each even isobar (as opposed to only one on each odd isobar). Of these two, the r-process produces only the stable nuclide with the most neutrons (see Figure 2-7). Since both nuclides are present in nature, we have not as yet reached the end of the element-synthesis story; there must be another process of considerable importance.

This remaining element-forming process also involves neutrons. As part of the smooth nuclear burn that characterizes most of a star's history, side reactions occur which release neutrons. As with the r-process, these neutrons build lighter elements into heavier elements. There is one major difference between the two processes; it has to do with the frequency of neutron hits. For the r-process, this frequency is extraordinarily high—so high, in fact, that even those nuclides with very short half-lives do not have a chance to undergo radiodecay before being hit again with a neutron. In contrast, the neutron bombardment associated with the steady nuclear fires of stellar cores is far more leisurely. Adequate time exists between hits for all but the radioisotopes with long half-lives to undergo radiodecay. Hence, the path followed by this so-called s-process follows the belt of stability (see Figure 2-8). This is the reason physicists refer to it as the s-process—for slow process. The s-process produces most of the stable nuclides not produced by the r-process.

There are a few nuclides that are not produced by either the s- or the r-process. These isotopes have abundances in nature

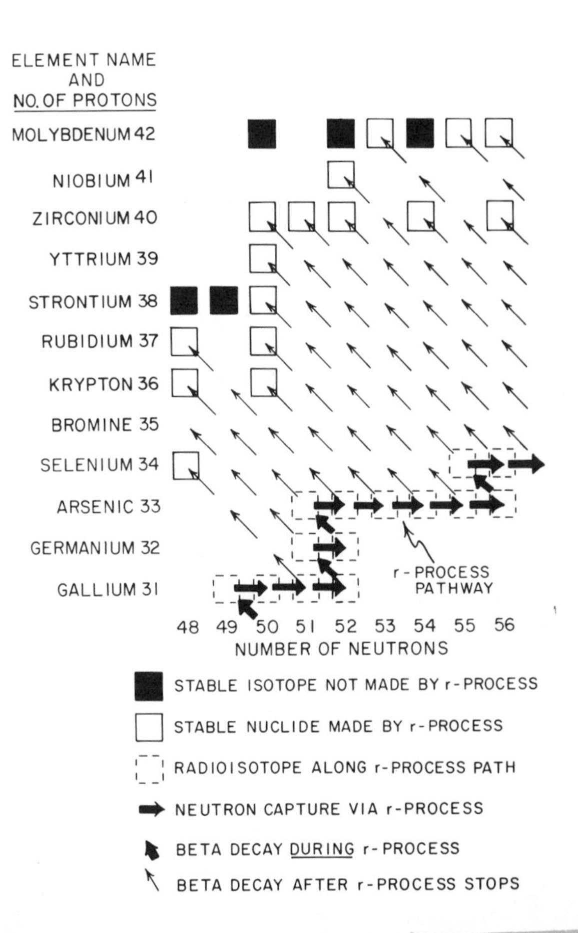

Figure 2-7. A segment of the r-process pathway: Rapid-fire neutron bombardment adds neutrons until a nuclide cannot hold any more. Only then does the nuclide undergo beta decay to become the next heavier element. This process—neutron capture to saturation followed by beta decay—is repeated over and over again, producing successively heavier elements. The r-process buildup occurs during the explosion that destroys the red giant. Hence it ends abruptly. The neutron flux stops and the highly radioactive isotopes on the r-process pathway emit beta particles one after another until stability is achieved. Note that in the case of those isobars for which two stable nuclides exist, only the neutron-rich nuclide of the pair is produced by the r-process.

about a hundred times lower than those of their r- and s-process brothers. They are produced by protons released as part of the side reactions occurring in stellar fires.

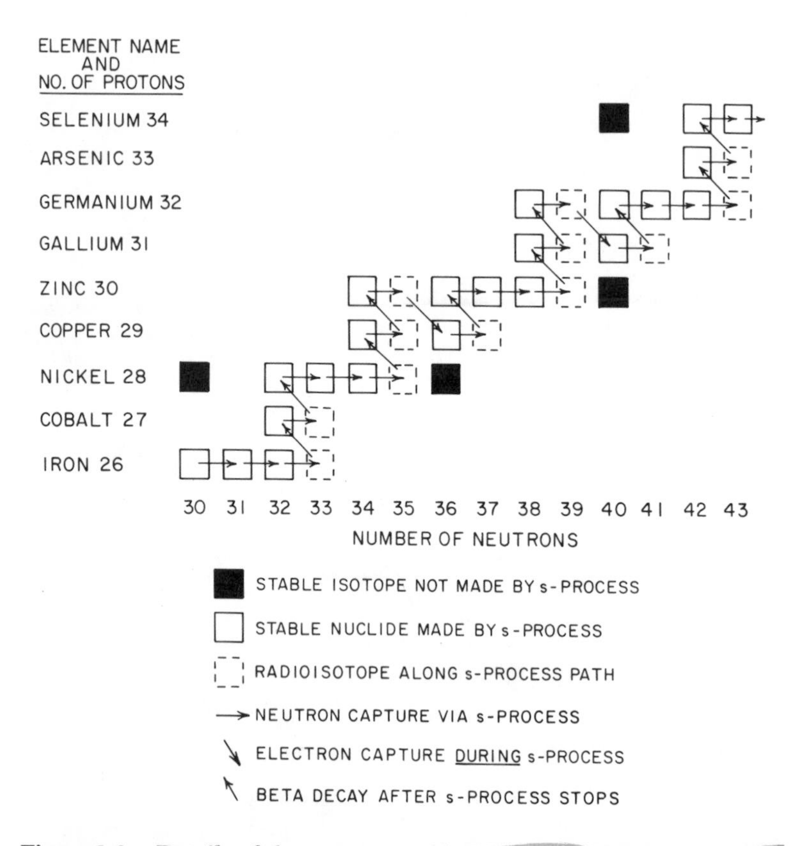

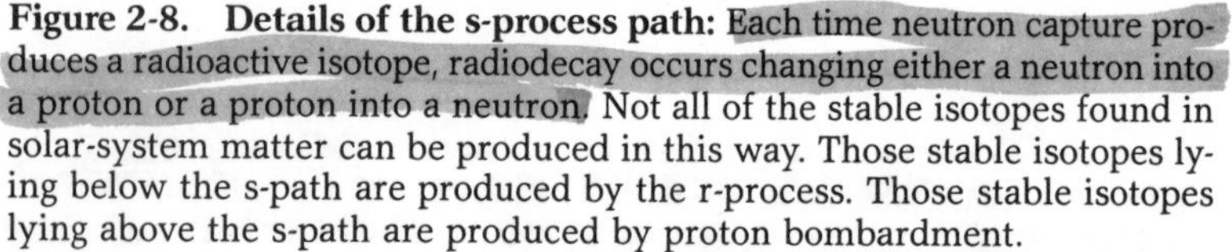

Figure 2-8. Details of the s-process path: Each time neutron capture produces a radioactive isotope, radiodecay occurs changing either a neutron into a proton or a proton into a neutron. Not all of the stable isotopes found in solar-system matter can be produced in this way. Those stable isotopes lying below the s-path are produced by the r-process. Those stable isotopes lying above the s-path are produced by proton bombardment.

Evidence Supporting the Stellar Hypothesis

Shall we accept the progressively intensifying nuclear fires and catastrophic explosions proposed by astrophysicists to explain the synthesis of the 90 elements heavier than helium? Can they be defended with hard evidence, or are they merely cosmic fairy tales? Clearly, no one has ever sent a probe into the core of a star. Thus, we have no direct evidence regarding these phenomena. Rather, our evidence is indirect. We have already mentioned two important lines of evidence. First, the only conceivable source of the energy needed to keep stars burning is nuclear. The pressures and temperatures computed for the cores of very large stars are adequate to permit not only hydrogen to burn but also helium and even heavier elements. Second, explosions of large stars have been observed (see Figure 2-9).

A third line of evidence comes from the element technetium, which is not present in the Earth. Nor are its dark lines present in sunlight spectra. The reason is that this element has no stable isotopes. However, the dark lines characteristic of the element technetium do appear in the spectra of the remnants of supernovae explosions. Technetium has two isotopes with moderately long half-lives: ^{97}Tc (2.6×10^6 years) and ^{98}Tc (4.2×10^6 years). These isotopes persist for millions of years after production during a supernova. They would have completely disappeared, however, during the 4.5×10^9 years that have passed since our solar system formed. Thus, the presence of the dark lines of technetium in the nebula produced by a supernova provides powerful support for the hypothesis that elements are being formed in red giants.

A fourth line of evidence comes from the relative abundances of the various kinds of nuclides in the Sun. Astrophysicists have carried out elaborate calculations designed to show what the proportions should be if elements were produced in red giants. These calculations reproduce very nicely the important features of the element-abundance curve. In Figure 2-10 the nuclide abundances are plotted as a function of the particle number. While there are many of the same features as in the abundance-versus-proton number curve in Figure 2-1, some additional ones appear.

While we cannot probe the reasons for all the features in this curve, a few are very important and easily grasped. One such feature is the peak associated with iron (at particle number 56 in Figure 2-10). If the stars that explode have cores of iron, then it is not surprising that the major nuclide of iron (^{56}Fe) is more abundant than its neighbors in universe material. One might ask, in fact, why the iron peak is not even more prominent. If all

Figure 2-9. Evidence for supernova explosions: Photographs taken before and after a supernova explosion.

June 1959

May 1972

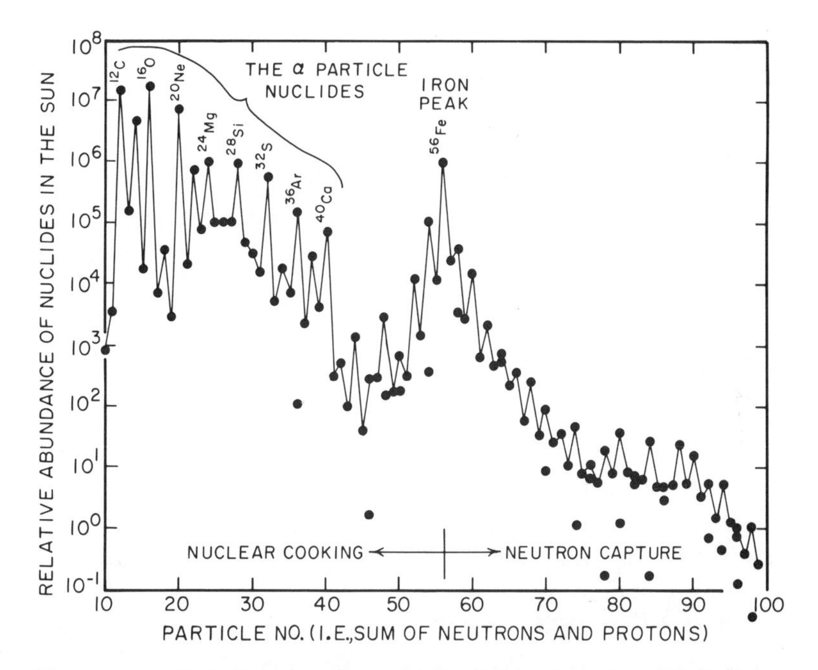

Figure 2-10. Relative abundances of individual nuclides: In the mass range 10 to 50, nuclides with particle numbers divisible by 4 (i.e., 12, 16, 20, 24, 28, 32 . . .) have abundances far above those of their neighbors. They are referred to as the *α*-particle nuclides. In the particle number range 50 to 100 the abundances of nuclides with an even particle number stand about a factor of 3 above those for their odd-numbered neighbors. Where more than one point is shown at a given mass number, two different nuclides with the same neutron-plus-proton number exist.

the material at the star's interior were converted to iron, then elements such as carbon, oxygen, magnesium, and silicon should be absent in the debris from supernovae. While this is the case for the star's core, it is not for the layers of gas surrounding the core. While forming iron in their cores, these mighty stars burn other nuclear fuels in layers around the core.

Another feature in the curve is the deep valley in the particle-number range from 6 to 11. We have already pointed out that no stable nuclides of mass 5 and 8 exist. The stable nuclides of the elements lithium (^{6}Li and ^{7}Li), beryllium (^{9}Be), and boron (^{10}B and ^{11}B) are all highly flammable in nuclear furnaces, and hence they are nearly absent in the debris thrown out by a supernova. Indeed, they are so flammable that the presence of any of the nuclides in universe matter is surprising to astrophysicists. It suggests that the small amount found may have formed in some other way.

Another feature of the abundances is the prominence in the mass range 10 to 40 of those nuclides with mass numbers divisible by four. Since they are aggregates of the very stable ^{4}He nucleus, they are the primary products of nuclear fires.

Enhanced Abundance for Low-Neutron-Capture Cross-Section Nuclides

For the nuclides formed by the s-process there is a relationship between the abundance of the nuclide and a property of the nuclide called its "neutron-capture cross section." Like targets in a shooting gallery some nuclei appear larger to neutron missiles than do others.As we see in Figure 2-11, those nuclides that present big targets for neutrons (those with large capture cross sections) have lower abundances than those nuclides that present small targets. This is exactly what is to be expected if these nuclides were formed by neutron irradiation of the slow variety.

This is best understood through an analogy. Imagine a county fair with a baseball throwing gallery. The participants attempt to throw balls through suspended rings. There are rings of two different sizes: big ones for small prizes and small ones for big prizes. As a gimmick to draw business away from his competitors, the proprietor devises a scheme which changes the size of a ring from big to small or from small to big when a ball passes through and lands in the mesh backing. Each day when he opens shop, the proprietor sets three-quarters of his rings with the big openings and one-quarter with the small openings. He notes that every night at closing time when he goes around to collect the balls from the mesh bags he finds considerably more of the rings in the small configuration. The proprietor eventually realizes that this is a logical consequence of his scheme. For the same reason that he gives out fewer big prizes than small prizes, he ends up

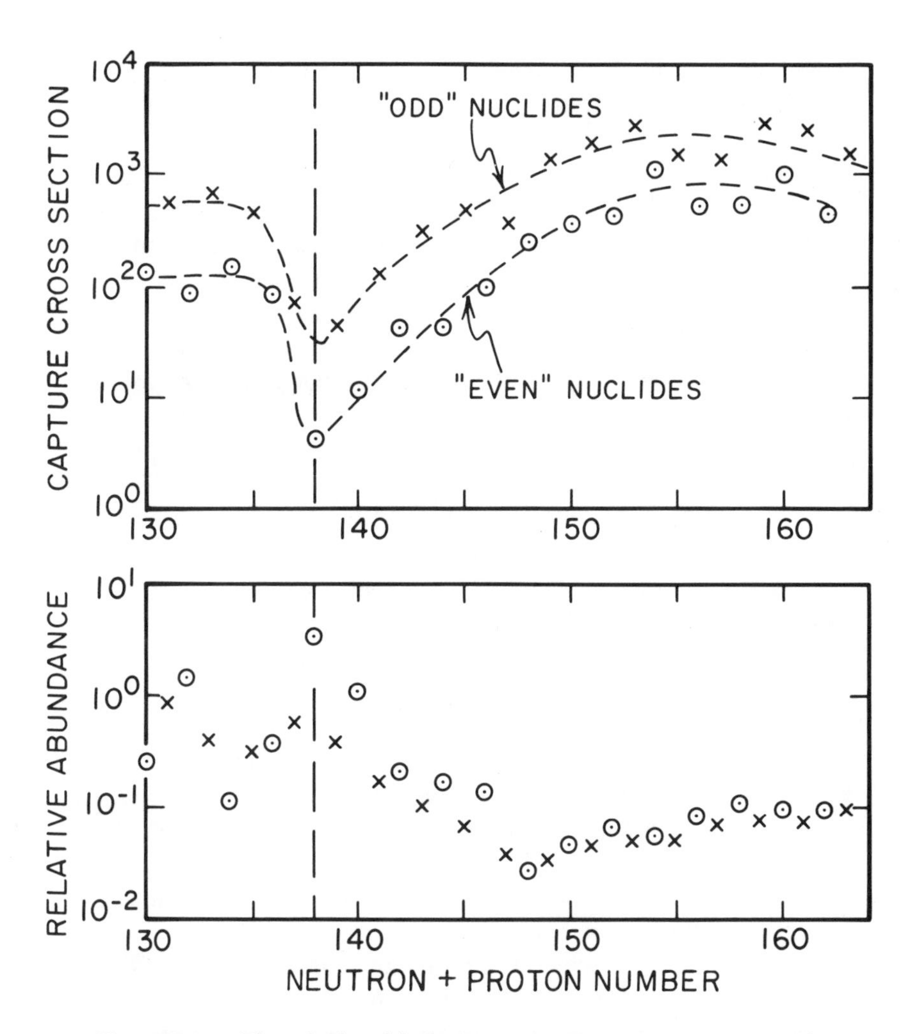

Figure 2-11. The relationship between neutron-capture cross sections and abundance: In the upper panel is shown the neutron-capture cross sections of nuclides produced by the s-process as a function of mass of the nuclide. Note the smoothness of the trend; note also that nuclides with an even number of nuclear particles have lower cross sections than those of their odd-numbered neighbors. Finally, note the minimum in the cross sections for both the even and odd nuclides near mass 138. Nuclides with this and neighboring masses have 82 neutrons, one of the magic numbers (see Fig. 2-6). As can be seen there is an inverse correlation between abundance and cross section. Nuclides with low capture cross section are higher in abundance.

with more rings in the small configuration. The probability that a ball thrown by his inaccurate clients will by chance pass through a small ring is much less than the probability that it will pass through a large ring!

The rings represent the nuclei being irradiated. The baseballs represent the neutrons. The size of the ring represents the neutron-capture cross section. Like a ring, each time an atom is hit by a neutron its capture cross section changes. Like the rings at the end of the night's activity, there are more atoms with small capture cross sections than atoms with large cross sections! Thus, the nuclides formed by this process have relative abundances in accord with expectation.

The dependence of nuclide abundance on neutron-capture cross section produces peaks in the abundance curve at particle numbers 138 and 208 (see Figure 2-12). These nuclides have what physicists refer to as magic neutron numbers. For some reason, nuclides with 82 and 126 neutrons are unusually stable. One way this stability expresses itself is in low cross sections for the capture of an additional neutron. Because of this, these nuclides were produced in greater abundance than their neighbors during the s-process.

As can be seen in Figure 2-12, in addition to the abundance peaks at nuclides with 138 and 208 particles, there are abundance peaks centered at particle numbers 130 and 194. The nuclides responsible for these peaks were produced by the r-process. They also owe their enhanced abundance to the magic neutron numbers 82 and 126. As can be seen in Figure 2-6, jogs in the r-process path occur at these magic neutron numbers. The reason is that nuclides with these ideal numbers of neutrons are loath to take on an additional one. Hence, there are several successive saturation-point nuclides at each of these magic neutron numbers. This creates steps in the r-process pathway. Furthermore, because of the enhanced stability of these nuclides, the pause before beta decay is a bit longer. This pause leads to an enhanced abundance for these nuclides. When the r-process onslaught ends, they undergo successive beta decays until they reach the stability belt. Thus, the stable nuclides formed in this

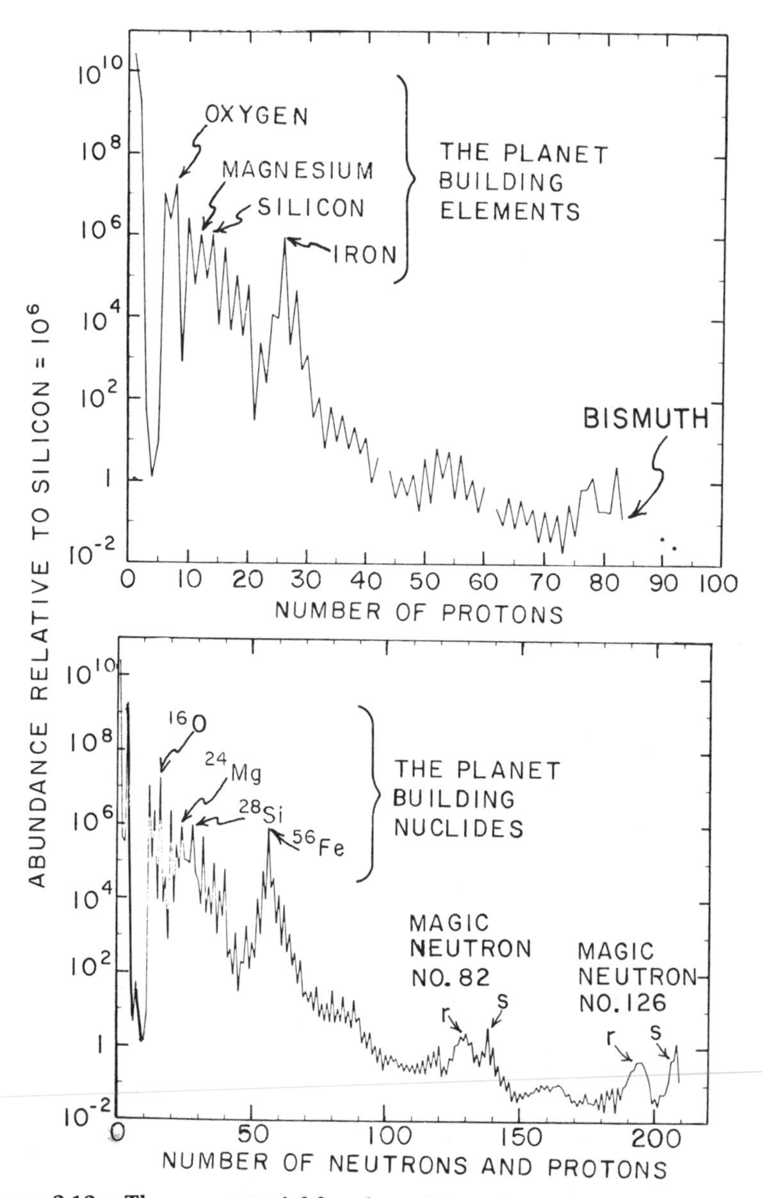

Figure 2-12. The raw material for planet formation: The upper diagram shows the relative abundances of the elements. Up to bismuth there are only two elements not found in nature; technetium (element 43) and promethium (element 61). The lower diagram shows the relative abundance, of the isobars. Only two isobars of nuclear number less than 208 are not represented in nature, those of mass 5 and of mass 8.

way have less than the magic number of neutrons. This is why the abundance peaks separated. The s-process peak is right at the magic neutron number. The r-process peak is displaced to a lower neutron number.

The Preference for Even Numbers

Even numbers have preference in universe matter. As can be seen from the chart of the nuclides (Figure 2-13), elements with an even number of protons generally have several isotopes, while elements with an odd number of protons have only one isotope. To understand how this comes about we must understand the diagrams in Figure 2-14. In each of the panels of this diagram are shown the masses for five nuclides which are isobaric brothers (that is, they all have the same number of nuclear particles). In each case the most stable nucleus is that with the smallest mass. Hence, of the nuclides shown for isobar 103, that with 58 neutrons and 45 protons is the most stable. For isobar 102 it is the nuclide with 58 neutrons and 44 protons. A nucleus can change one of its protons into a neutron by capturing an electron, or it can change a neutron into a proton by beta decay. These transformations occur, however, only if they lead to a decrease in mass.

With these general principles in mind, consider the shapes of the two curves in Figure 2-14. That for isobar 103 has a simple U shape. Were a ball released on either side of a trough of this shape it would come to rest on the bottom. In nuclide terms this means that no matter what nuclide of particle number 103 is manufactured, it will undergo successive radiodecays until it achieves the 45-proton–58-neutron combination. Thus, only one stable nuclide with 103 particles exists. The simple shape of the curves for this and other odd isobars has to do with the fact that all the combinations are either odd–even or even–odd. As there is no particular stability difference between odd–even and even–odd nuclides, the mass trend is smooth. The position of the valley bottom corresponds to the most favorable ratio of neutrons to protons.

The curve for isobar 102 is not regular. A peak sits in the middle

Figure 2-13. Chart of the nuclides: Shown in this series of diagrams are all the nuclides present in nature. The black squares represent radioactive isotopes. Some of these are long-lived remnants of element production in stars. Others are being produced in very small quantities by cosmic rays bombarding our atmosphere. To avoid confusion, the decay chains of long-lived thorium and uranium isotopes are shown separately.

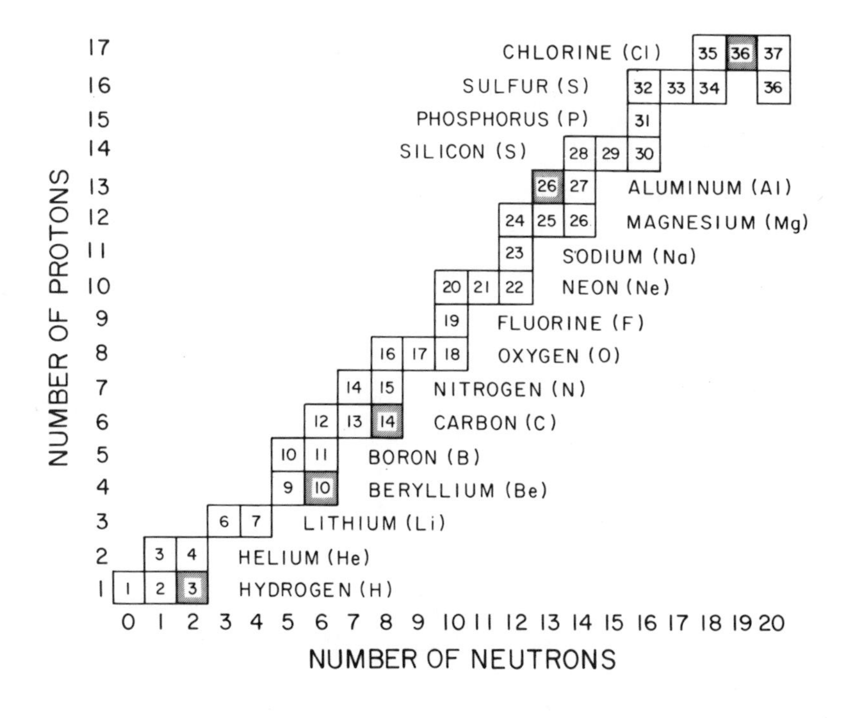

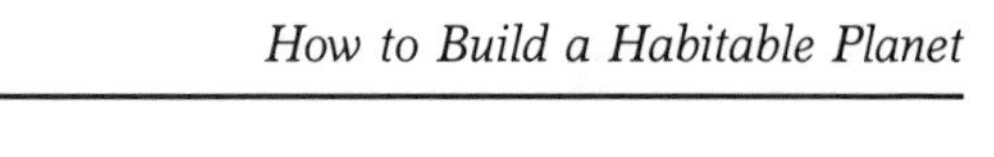

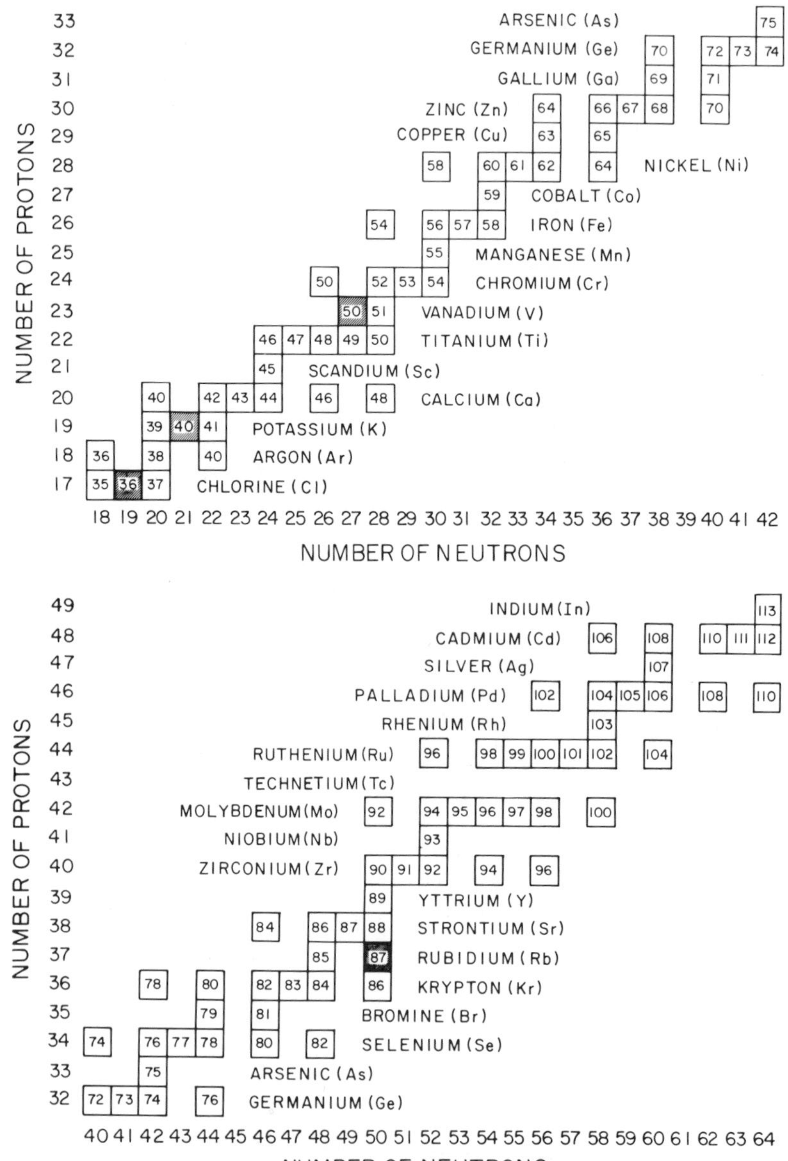
ARSENIC (As)
GERMANIUM (Ge)
GALLIUM (Ga)
ZINC (Zn)
COPPER (Cu)
NICKEL (Ni)
COBALT (Co)
IRON (Fe)
MANGANESE (Mn)
CHROMIUM (Cr)
VANADIUM (V)
TITANIUM (Ti)
SCANDIUM (Sc)
CALCIUM (Ca)
POTASSIUM (K)
ARGON (Ar)
CHLORINE (Cl)
NUMBER OF PROTONS
NUMBER OF NEUTRONS
INDIUM (In)
CADMIUM (Cd)
SILVER (Ag)
PALLADIUM (Pd)
RHENIUM (Rh)
RUTHENIUM (Ru)
TECHNETIUM (Tc)
MOLYBDENUM (Mo)
NIOBIUM (Nb)
ZIRCONIUM (Zr)
YTTRIUM (Y)
STRONTIUM (Sr)
RUBIDIUM (Rb)
KRYPTON (Kr)
BROMINE (Br)
SELENIUM (Se)
ARSENIC (As)
GERMANIUM (Ge)
NUMBER OF PROTONS
NUMBER OF NEUTRONS

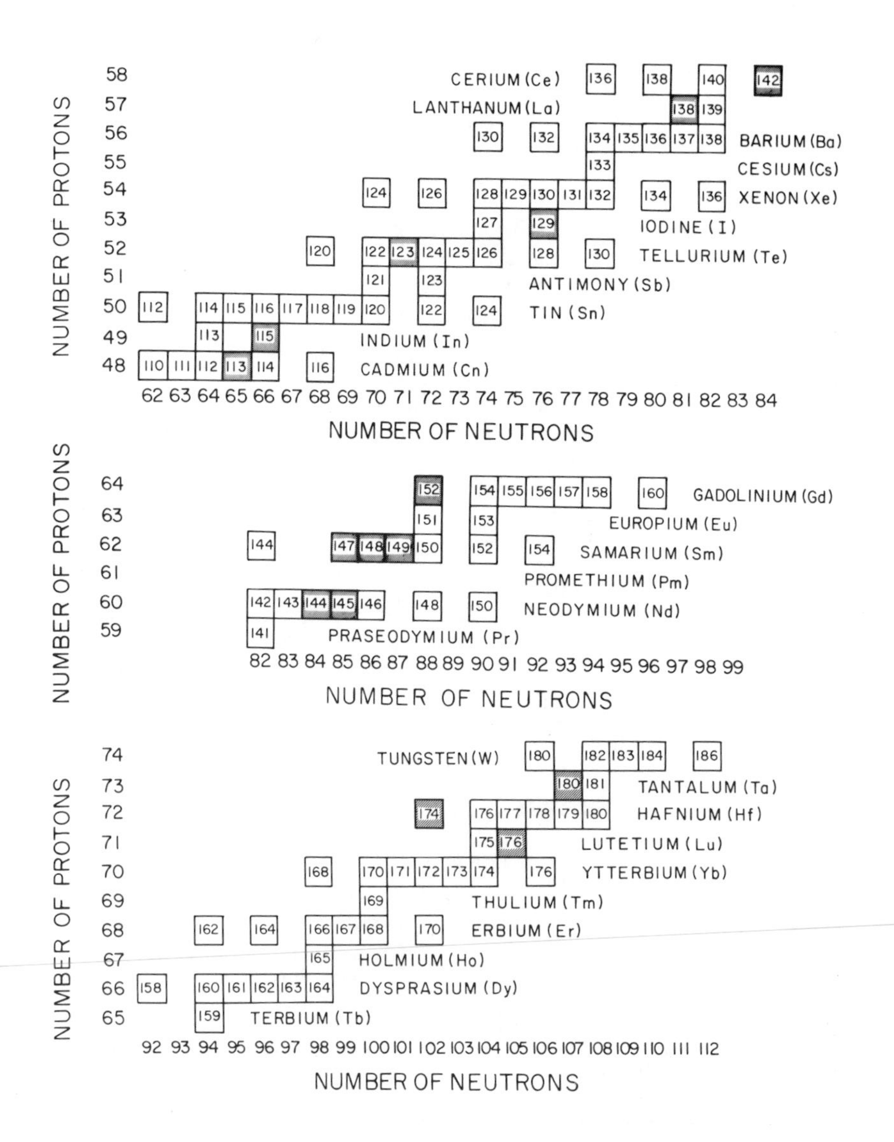
NUMBER OF PROTONS
CERIUM (Ce)
LANTHANUM (La)
BARIUM (Ba)
CESIUM (Cs)
XENON (Xe)
IODINE (I)
TELLURIUM (Te)
ANTIMONY (Sb)
TIN (Sn)
INDIUM (In)
CADMIUM (Cn)
NUMBER OF NEUTRONS
NUMBER OF PROTONS
GADOLINIUM (Gd)
EUROPIUM (Eu)
SAMARIUM (Sm)
PROMETHIUM (Pm)
NEODYMIUM (Nd)
PRASEODYMIUM (Pr)
NUMBER OF NEUTRONS
NUMBER OF PROTONS
TUNGSTEN (W)
TANTALUM (Ta)
HAFNIUM (Hf)
LUTETIUM (Lu)
YTTERBIUM (Yb)
THULIUM (Tm)
ERBIUM (Er)
HOLMIUM (Ho)
DYSPRASIUM (Dy)
TERBIUM (Tb)
NUMBER OF NEUTRONS

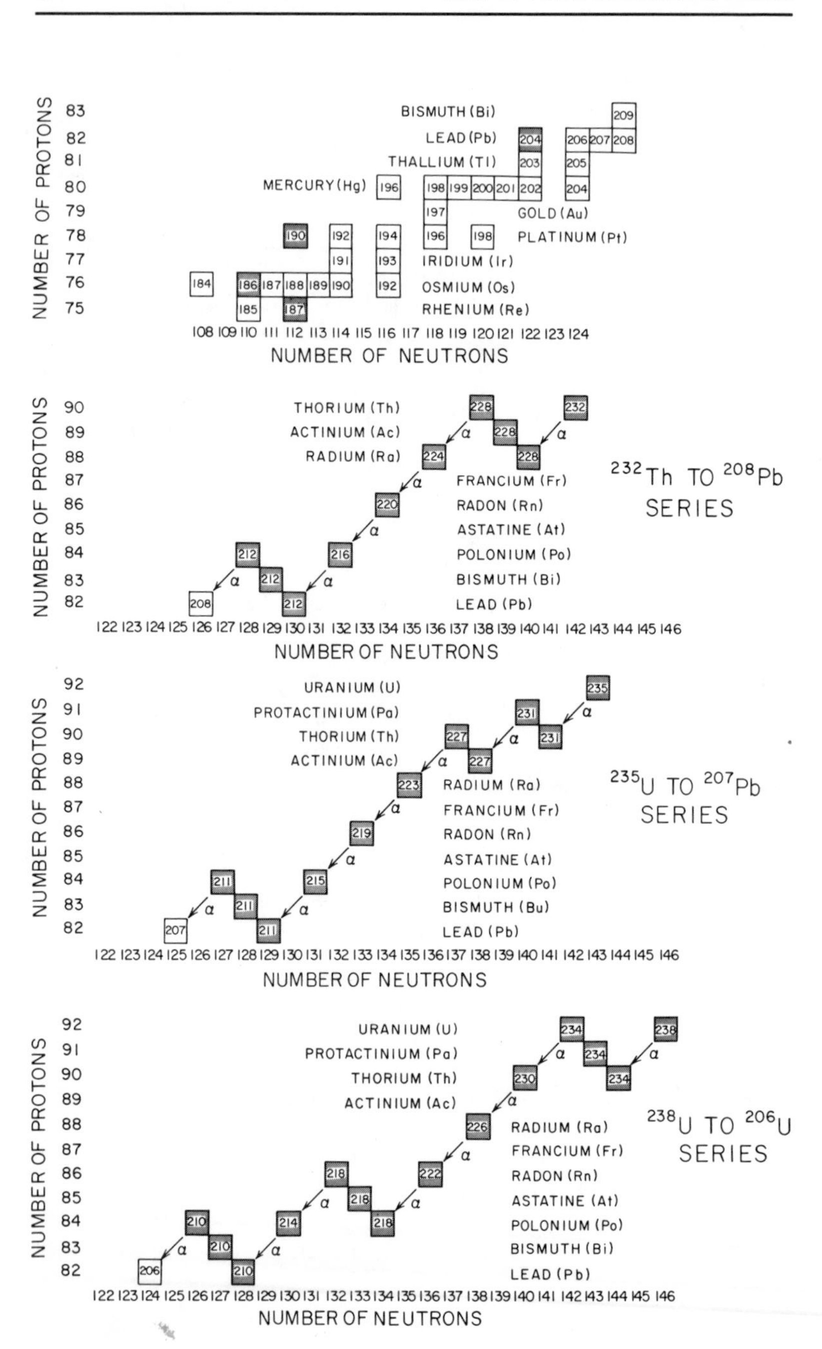
NUMBER OF PROTONS
NUMBER OF NEUTRONS
BISMUTH (Bi)
LEAD (Pb)
THALLIUM (Tl)
MERCURY (Hg)
GOLD (Au)
PLATINUM (Pt)
IRIDIUM (Ir)
OSMIUM (Os)
RHENIUM (Re)
THORIUM (Th)
ACTINIUM (Ac)
RADIUM (Ra)
FRANCIUM (Fr)
RADON (Rn)
ASTATINE (At)
POLONIUM (Po)
BISMUTH (Bi)
LEAD (Pb)
^{232}Th TO ^{208}Pb SERIES
URANIUM (U)
PROTACTINIUM (Pa)
THORIUM (Th)
ACTINIUM (Ac)
RADIUM (Ra)
FRANCIUM (Fr)
RADON (Rn)
ASTATINE (At)
POLONIUM (Po)
BISMUTH (Bu)
LEAD (Pb)
^{235}U TO ^{207}Pb SERIES
URANIUM (U)
PROTACTINIUM (Pa)
THORIUM (Th)
ACTINIUM (Ac)
RADIUM (Ra)
FRANCIUM (Fr)
RADON (Rn)
ASTATINE (At)
POLONIUM (Po)
BISMUTH (Bi)
LEAD (Pb)
^{238}U TO ^{206}U SERIES

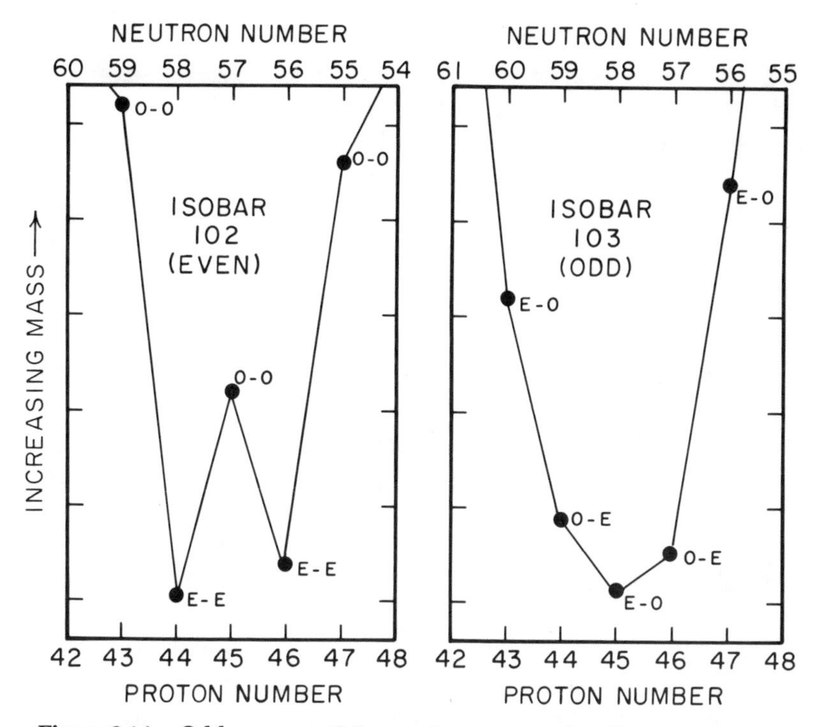

Figure 2-14. Odd–even particle number systematics: Shown here are the masses for two sets of nuclides. On the left are shown nuclides of particle number 102 (an even number). On the right are shown nuclides of particle number 103 (an odd number). The smaller the mass the more strongly the nuclide is bound together. For odd isobars, of the various possible neutron–proton combinations, one always has a lower mass than both its adjacent neighbors. For even isobars, there are usually two such nuclides. The reason for this difference is that odd-odd neturon–proton combinations are less tightly bound than even–even combinations.

of the trough. A ball released on the right side of a trough shaped like this would come to rest in the valley to the right of the peak; a ball released on the left would come to rest in the valley to the left side of the peak. The reason for this rugged topography is that the neutron–proton combinations are either both odd or both even. The even–even combinations are considerably more stable (hence, less massive) than the odd–odd combinations. This situation permits two stable nuclides to exist for isobar 102. One is a nuclide with 44 protons and 58 neutrons and the other a nuclide with 46 protons and 56 neutrons.

This preference of nature for even–even combinations destines odd-numbered elements to one isotope and gives the even-numbered ones several. It is for this reason that the element-abundance curve shown in Figure 2-12 has its strong sawtoothed character.

Summary

Clearly, the Sun cannot have been among the first stars to form in the universe. Were this the case, it would contain only hydrogen and helium. Instead, the Sun must be a latecomer. Its element composition is the product of the explosive deaths of large numbers of red-giant stars which formed and died long before the advent of our Sun. Thus, the Sun's origin had to await not only the formation of our galaxy but also the origin and death of myriads of red giants within our galaxy.

From what we have learned here, heavy-element formation must have gone on in all the galaxies making up our universe. The ingredients for rocky planets have long been available everywhere in the universe. Hence, the generation of Earth-like planets has not been impeded by a lack of raw material!

Supplementary Readings

The Origin of the Chemical Elements , *by R.J. Taylor, 1972, Wykeham Publications, Ltd.*

A book summarizing the estimates of the abundances and theories of the origin of the elements making up our universe.

Essays in Nuclear Astrophysics, *edited by C.A. Barnes, D.D. Clayton and D.N. Schramm, 1982, Cambridge University Press.*

A series of papers written for specialists summarizing the status of knowledge regarding nuclear processes in stars as of 1982.

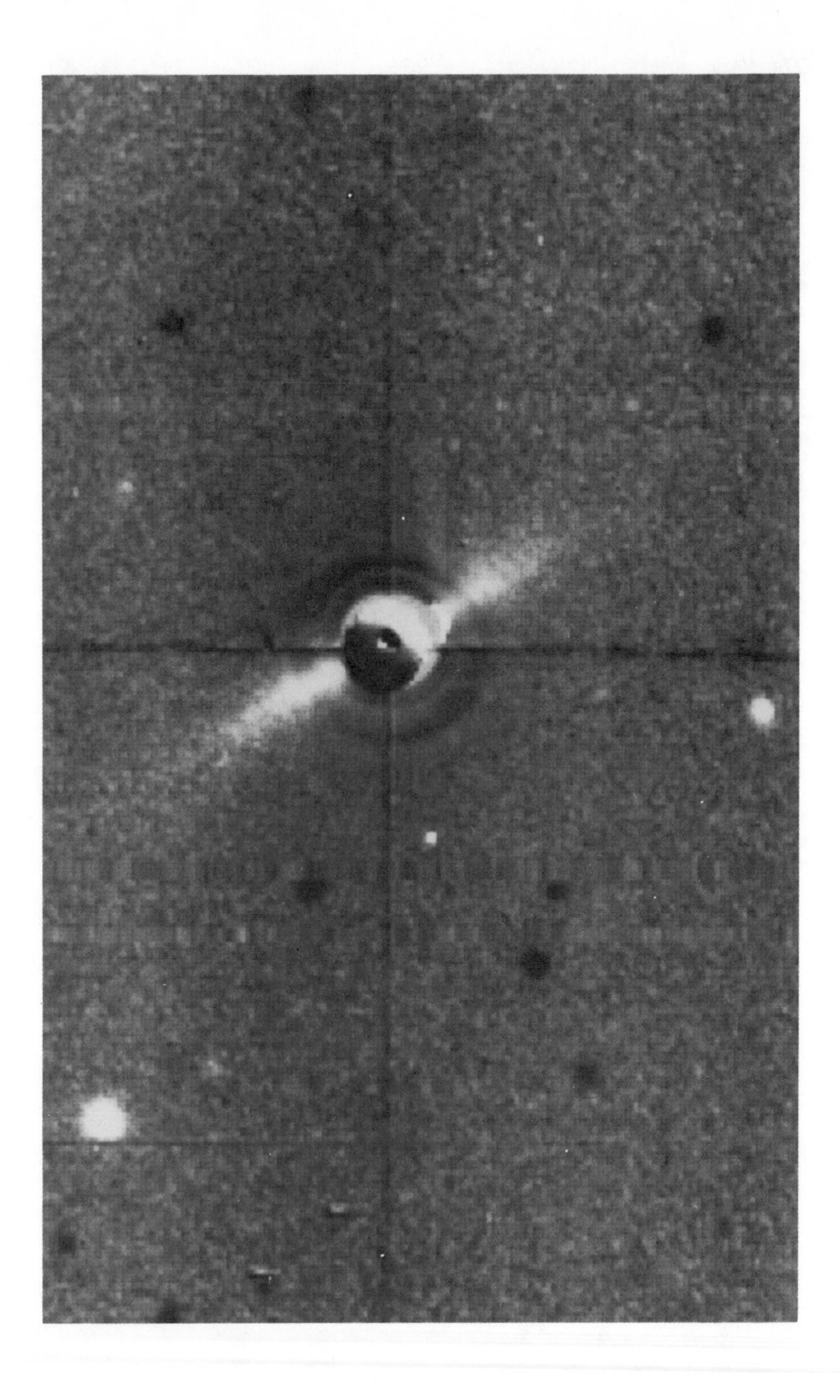

Beta Pictoris

Chapter Three

The Heavy Construction:

The Formation of Planets

Our Sun is thought to have formed from a small cloud of gas and dust which, after lingering for billions of years in one of the Milky Way's spiral arms, succumbed to its own gravitational pull and collapsed. Most of the material in this cloud was drawn into a central body. Thus, our Sun has a composition nearly identical to that of the original cloud. It contains 99 percent hydrogen and helium and about 1 percent the remaining 90 elements.

A small amount of the matter in the cloud ended up in a nebular disk around the newly formed Sun. This material somehow aggregated into planets, moons, asteroids, and comets. The composition of these objects is very different from that of the Sun. This difference is the result of a gigantic chemical separation: those elements contained in dust grains and snowflakes were largely retained through incorporation into the objects making up the planetary system; those elements in gaseous form were largely driven off by the particles streaming forth from the Sun.

In those parts of the disk close to the hot Sun only the least volatile of the chemical compounds were in solid form. In the most distant reaches of the disk, all of the elements except hydrogen and helium were in solid form. Because of this, the comets and planets of the outer solar system have a chemical composition quite different from that of the asteroids and planets of the inner solar system.

While we have some sense of what may have taken place during the birth of our planetary system, the details are mostly lacking. The problem is that we have yet to see planet formation in progress around some neighboring newly born star. The closest we have come is a picture recently provided by a new device called a charge-coupled sensor which shows that the star Beta Pictoris is surrounded by a disk. Astronomers surmise that within this disk a planetary system may be forming.

Introduction

While there is no doubt that the Earth and its fellow planets are by-products of the formation of the Sun, the mechanisms that allowed these objects to come into being remain obscure. Several competing hypotheses exist. None, however, ranks more than 4 on our 1-to-10 scale of credibility. As we shall see, this gap in our knowledge proves to be a stumbling block in any attempt to assess the likelihood of producing a habitable planet. The extent to which a planet is endowed with several elements important to its surface environment is closely tied to details of the formation process. Since we have not even been able to get the first-order aspects of the formation process quite straight, these details lie beyond our grasp.

While not having enough information to pin down the exact manner in which our planetary system formed, we do have some extremely important clues. These clues come from four sources:

1) the orbits, volume, and mass* of each of the nine planets;
2) chemical measurements made by unmanned space probes;
3) direct chemical analyses of rocks from the Earth;
4) the textures and chemical compositions of meteorites.

Planetary Vital Statistics

By tracking the motions of the planets, astronomers have been able to show that their orbits are nearly circular and lie nearly in the same plane (see Table 3-1). This plane corresponds to the Sun's equator. All the planets orbit the Sun in the same sense.

*The mass of an object is a measure of the number of protons and neutrons it contains. A more familiar term is weight; the weight of an object is the gravitational pull exerted on it by the Earth (or any other planet or moon on which it is situated). Were an astronaut to have taken his bathroom scale with him to the Moon he would have found that he weighed only one-sixth as much there as here. His mass, however, would not be changed.

Table 3-1. Characteristics of the orbits of the nine planets and of the largest object in the asteroid belt:

Planet	Radius of orbit (10^{13} centimeters)	Time for one revolution about Sun (years)	Inclination of orbit* (degrees)	Eccentricity of orbit†	Length of day (days)
Mercury	0.58	0.24	7.0	0.21	59
Venus	1.08	0.62	3.4	0.01	243
Earth	1.50	1.00	0.0	0.02	1.0
Mars	2.29	1.88	1.9	0.09	1.0
Asteroid Ceres	4.15	4.60	10.6	0.08	0.4
Jupiter	7.70	11.9	1.3	0.05	0.4
Saturn	14.3	29.5	2.5	0.06	0.4
Uranus	28.3	84.0	0.8	0.05	1.0
Neptune	45.1	164.8	1.8	0.01	0.9
Pluto	59.2	247.7	17.2	0.25	6.2

*The plane of the Earth's orbit is used as the reference.
†A measure of the deviation from circularity.

This direction is the same as the direction of the Sun's spin. This basic organization is consistent with the hypothesis that the cloud from which the Sun formed was spinning. The spin of the Sun and the orbits of the nine planets conform to the axis about which the original cloud turned. Were the orbits, instead, helter-skelter, astronomers might have been tempted to conclude that the planets formed elsewhere and were subsequently captured by the Sun.

Using the surveying method discussed in Chapter One, astronomers have been able to determine the distance from the Sun to each of the planets (see Table 3-1). If the asteroids orbiting in a belt between Mars and Jupiter are considered to be the remnants of an abortive attempt to form a tenth planet, then the spacing between the planets shows a certain degree of regularity. As shown in Table 3-2, moving out from the Sun, the distance between orbits of successive planets increases by roughly a factor of 1.6. While a number of early theories of planet formation pondered this regularity in spacing, little recent use has been made of this observation. It is neither so surprising nor so perfect as to guide us to a particular mechanism of planet formation.

Table 3-2. Regularity in orbit spacing: While the planets are spaced such that the radius of each orbit is roughly 1.6 times greater than that for the next nearest to the Sun, this relationship is far from perfect!

Planet	Actual Distance from Sun 10^{13}cm	Distance if the times 1.6 relationship were exact 10^{13}cm
Mercury	0.58	0.9*
Venus	1.08	1.4
Earth	1.50	2.2
Mars	2.29	3.6
Asteroids	4.05	5.7
Jupiter	7.80	9.2
Saturn	14.3	15
Uranus	28.8	23
Neptune	45.1	38
Pluto	59.2	60

*Adjusted so as to achieve the best match with the actual distances for the other eight planets.

The mass of a planet is determined from the gravitational influence it exerts on the orbits of its moons, on other planets, and on the space probes sent out from Earth. For all except Pluto, the tiniest and most distant planet, these masses are well established. The primary method used is a very clever one. It is based on a law established during the early part of the 17th century by Kepler, a German astronomer and mathematician. This law provides a relationship between the distance at which a moon orbits its host planet (the orbital radius), the time it takes the moon to make one revolution about its host planet (the "period"), the mass of the host planet, and the universal constant of gravitation. It is as follows:

$$\text{period} = \text{universal constant of gravitation} \left(\frac{(\text{distance from host})^3}{\text{mass of host}}\right)^{1/2}$$

This relationship provided astronomers with a ready means to determine the masses of those planets that have captive moons (Mars, Jupiter, Saturn, Uranus, and Neptune). By observing both the time required for the moon to circle the planet and the distance of the moon from the planet they could calculate the host planet's mass. It is important to note that the period of a moon is independent of the moon's mass. This is why astronauts need not be afraid to leave their space ships or to let go of their wrenches. All objects orbiting at the same distance from a planet have the same period.

For those planets not having moons (Mercury, Venus, and Pluto), the masses had to be calculated by a more elaborate scheme. The orbit of each planet is influenced by the gravitational interactions with neighboring planets. Since the magnitudes of these perturbations depend on the masses of the neighboring planets, it is possible—from careful observations of the positions of the planets over a period of time and from laborious calculations—to extract the desired mass information. This procedure worked well for Mercury and Mars but not for tiny and distant Pluto.

The planetary masses change greatly from planet to planet (see Table 3-3). Jupiter and Saturn are giants. Neptune and Uranus, while somewhat less massive, are still far more massive than any of the four inner planets. Venus and Earth come next; they are remarkably similar in mass. Mars and Mercury are the least massive planets. The combined mass of all the objects in the asteroid belt is far less than the mass of Mercury. So we see that the process responsible for the building of planets led to objects differing widely in size. Furthermore, this variation in size bears no simple relationship to distance from the Sun.

One of the most difficult properties of the solar system to explain is the distribution of angular momentum.* A measure of this somewhat obscure quantity is the energy that would be transferred to a set of giant brake shoes designed to bring the planets to a halt in their orbits and to bring the slowly spinning Sun to a stop. While the Sun has more than 99.9 percent of the solar system's total mass, it has only 2 percent of the total angular momentum. The solar system derived its angular momentum from the slow spin of the gas cloud from which it originated.

---

*The angular momentum of a planet is given by the product of its mass, its orbital radius, and its orbital velocity.

Thus, as particles were drawn closer to the center of the system during the collapse of this cloud, their velocity increased (just as a skater's velocity increases as her arms and legs are drawn closer to her body). The problem is how the momentum associated with the particles that became part of the Sun was so efficiently transferred to the planets. Historically, this has proven to be one of the major stumbling blocks associated with scenarios of planet origin.

Density: A Guide to Chemical Composition

More revealing to hypotheses regarding the mode of origin of the planets is the ratio of the mass of a planet to its volume. This ratio provides a measure of the bulk density of the matter making up the planet, which in turn provides strong clues regarding the chemical composition of the planet. To understand this let us consider the relationship between the density and chemical composition of substances found on the Earth's surface. This relationship exists because atoms are far more similar in size than they are in mass. Atoms have diameters falling into a narrow range (1×10^{-8} to 4×10^{-8} cm). By contrast, the mass of atoms covers a far larger range. Hydrogen atoms have only one nuclear particle; uranium atoms have 238 nuclear particles. Thus, to a rough approximation, the heavier an element, the greater the density of the substances the element makes. Water, for example, has a bulk density of 1.0 grams per cubic centimeter. Each water molecule has two hydrogen atoms (one nuclear particle each) and one oxygen atom (16 nuclear particles). The total number of nuclear particles in a water molecule is 18. Hence, water has on the average six nuclear particles per atom. The mineral periclase has one magnesium atom (24 nuclear particles) for each oxygen atom (16 nuclear particles). Hence, periclase has on the average 20 nuclear particles per atom (3.3 times the number for water). Periclase has a density of 2.5 grams per cubic centimeter (2.5 times the density of water). Iron metal has a density of 7.5 grams per cubic centimeter (7.5 times that of water). Each iron atom has 56 nuclear particles (9 times as many as the average for the atoms in water).

Table 3-3. Characteristics of the planets:*

Planet Name	Radius 10^8 cm	Volume 10^{26} cm³	Mass 10^{27} gm	Density gm/cm³	Corrected density† gm/cm³
Mercury	2.44	0.61	0.33	5.42	5.4
Venus	6.05	9.3	4.9	5.25	4.3
Earth	6.38	10.9	6.0	5.52	4.3
Mars	3.40	1.6	0.64	3.94	3.7
Jupiter	71.90	15,560	1900	1.31	<1.3
Saturn	60.20	9130	570	0.69	<0.7
Uranus	25.40	690	88	1.31	<1.3
Neptune	24.75	635	103	1.67	<1.7
Pluto	1.6	0.17	?	?	?

*The mass of the Sun is 1.99×10^{33}gm, 1000 times the mass of Jupiter.
†Density a planet would have in the absence of gravitational squeezing.

Other examples are given in Table 3-4. While not exact, the proportionality between the density of a substance and the average number of nuclear particles per atom is reasonably good. Hence, the density of a planet has something to tell us about its chemical composition.

The situation is complicated a bit by the fact that the average density of a planet depends on its mass. The more massive the planet, the stronger its gravitational pull. The pressure generated by this gravity causes the planet's constituent atoms to contract in size. The greater the planet's mass, the greater its gravity and the greater the contraction. These gravitationally induced increases in bulk density are substantial even for relatively small planets like Earth.

Through studies employing laboratory devices capable of duplicating the pressures that exist deep within the Earth, geophysicists have been able to assess the influence of pressure on the density of materials of interest (rocks, iron, etc.). Using this information they have been able to determine how much the density of each of the four terrestrial planets has been increased by gravitational squeezing. These corrected densities are given in Table 3-3.

Table 3-4. The relationship between the density of a substance and the average number of nuclear particles in its constituent atoms:

Substance	Formula	Nuclear particles Atom	Density gm/cm³	Ratio*
Water	H_2O	6.0	1.00	6.0
Calcite	$CaCO_3$	20.0	2.72	7.4
Quartz	SiO_2	20.0	2.65	7.5
Gypsum	$CaSO_4 2H_2O$	14.3	2.32	6.2
Olivine	Mg_2SiO_4	18.3	3.20	5.7
Hematite	Fe_2O_3	32.0	5.26	6.1
Magnetite	Fe_3O_4	33.1	5.18	6.4
Diamond	C	12.0	3.50	3.4
Iron	Fe	56	7.50	7.5
Gold	Au	197	17.10	11.6

*Nuclear particles per atom divided by density. Were the relationship between the average number of nuclear particles per atom and density perfect, this ratio would be exactly the same for all ten compounds. For 8 of the 10 compounds the range is small (5.7 to 7.5). For gold and for diamond the deviation from the mean for the other 8 is sizable (a factor of about 2).

Corrected densities are not given for the major planets. The reason is that, due to their extremely high gravity, the corrections remain uncertain. Obviously, these corrected densities would be lower than the uncorrected values given in Table 3-3. Clearly, the four major planets must be made of elements with a smaller number of nuclear particles than the four terrestrial planets!

Beyond this major difference between the two groups of planets we see that there are also significant differences among the planets within each group. Of particular interest to us are the differences among the densities of the four terrestrial planets. As can be seen in Table 3-3, after correction for gravitational compression, we find that the matter in Mercury has the highest density and that in Mars the smallest density. Earth and Venus have nearly the same corrected density. These densities lie midway between that for Mercury and that for Mars.

While the gravity-corrected bulk densities give us an estimate of the average number of nuclear particles contained by the atoms of a given planet, there are a wide variety of combinations of elements through which this mean could be achieved. In order to determine which of these possible combinations is the correct one, we must have some additional information. The situation is similar to that faced by a woman about to open a birthday present. From the heft of the box she can eliminate many possibilities. If the box is quite heavy for its size, it could be a book or a tape recorder. If it is quite light for its size, it could be a slip or stockings. While providing a hot clue, the heft leaves many possibilities!

Chemical Clues from Meteorites

Our best evidence regarding the chemical composition of the terrestrial planets comes from pieces of rock that fall from the sky. These objects are called meteorites. Although most of these objects are pieces broken off of asteroids during their collisions with one another, a few have been shown to be hunks of rock blasted off the surface of the Moon and one or two are suspected of being hunks of rock blasted off the surface of Mars (by the impacts of large objects from the asteroid belt).

One might ask why we must turn to these obscure objects when we have at our disposal from the Earth's surface as many rocks as we could possibly want. The answer is that the Earth is chemically layered. The material at the surface has a chemical composition atypical of that for the planet as a whole. Earth surface rocks have densities in the range of 2.6 to 3.0 grams per cubic centimeter. By contrast, the gravitation-corrected bulk density of average Earth material is about 4.3 grams per cubic centimeter. So we look to these objects that fall upon the Earth with the hope that they are less biased samples of the material that makes up the bulk of the terrestrial planets.

Our collections of meteorites come not only from objects whose fiery passages through the atmosphere were seen by man but also from finds of similar objects that fell in the past. The mother lode of these finds is the Antarctic ice cap. In certain places on this cap, wind scour and evaporation remove more material than falls as snow. Any meteorites contained in this disappearing ice are left on the surface as a lag, just as pebbles are left behind in desert areas whose sands are being swept away by strong winds. Over the last decade thousands of meteorites have been recovered from Antarctica by expeditions of American and Japanese scientists.

Most meteorites are made of stony material. Some of these stony meteorites contain millimeter-sized spheres called chondrules (see Figure 3-1). These chondrules are unique to meteorites; they have never been found in the rocks making up the surface of the Earth. Much attention has been given to these little spheres, and the conclusion is nearly universal. They were once molten droplets. Some think these drops were formed like raindrops as a hot nebular gas cooled; some think they were formed when small fragments of nebular dust were heated beyond their melting point; some think they were formed by splashes created by impacts on the surfaces of the growing asteroids; and some think they were fused by bolts of lightning in the early nebula. No one, however, has conceived a reasonable way to manufacture these spheres inside a planet. Thus, meteorites containing chondrules must consist of material that formed in the solar nebula. To become chemically segregated the material in a planet must melt. Clearly, the meteorites bearing chondrules have never melted. Thus, these objects are likely to carry the chemical information we seek.

Chondrule-bearing meteorites appear to have been formed through the aggregation of grains formed under quite different conditions. As the chondrules themselves were once liquid, they formed at high temperatures ($\simeq$1300° C). In a rare class of meteorites called the carbonaceous chondrites these chondrules coexist with minerals that are unstable above 100° C. Hence, these fragments formed in a low-temperature environment. Other types of chondrule-bearing meteorites appear to have been baked at temperatures high enough to destroy any low-temperature grains they might once have contained. As this baking surely caused the partial loss of volatile prone elements, in the quest for the best estimate of the composition of the primitive dust grains in the solar nebula attention has focused on carbonaceous chondrites.

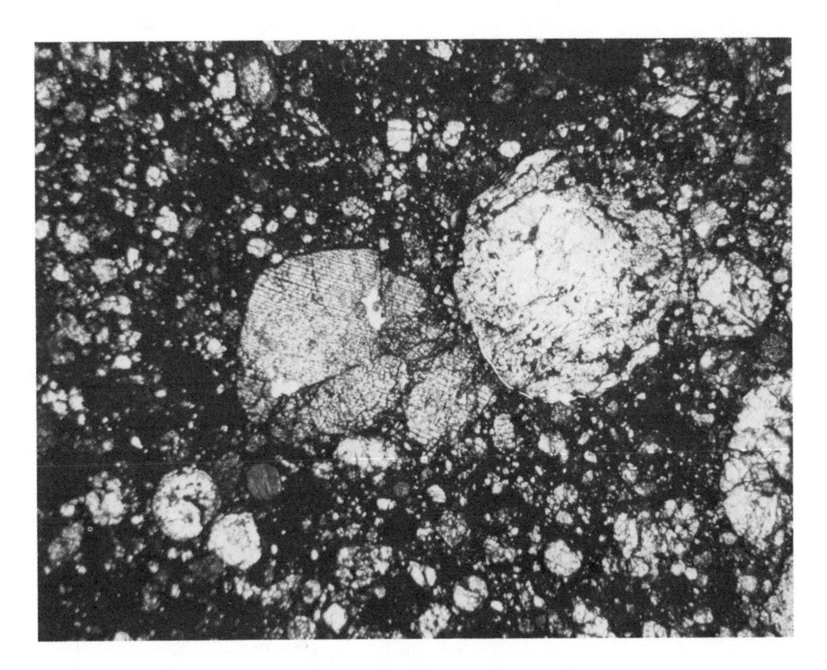

Figure 3-1. Photomicrograph of chondrules from a stony meteorite: The chondrules are about 1 millimeter in diameter.

The chemical composition of carbonaceous chondrites is compared to that for the Sun in Figure 3-2. Included in this figure are only elements of moderate to low volatility. The agreement is remarkable. Not only does this agreement lend confidence to the spectral method of chemical analysis of stars, but it also fortifies the conclusion that carbonaceous chondrites provide a chemically unbiased sample of the solar system's less volatile elements.

The most dramatic feature of the average chemical composition of all chondrites (carbonaceous and ordinary) is the dominance of four elements: oxygen, silicon, magnesium, and iron. As summarized in Table 3-5, magnesium, silicon, and iron constitute 91 percent of the metals present in ordinary chondrites. Four elements—aluminum, calcium, nickel, and sodium—form a group that runs a poor second, and six other elements form a group that runs an even poorer third.

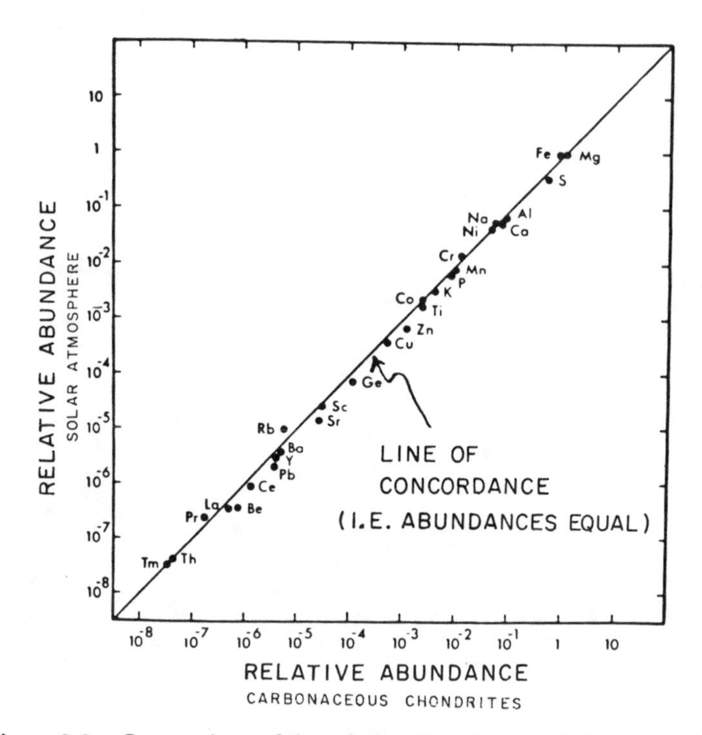

Figure 3-2. Comparison of the relative abundances of elements of low and moderate volatility in the Sun's atmosphere with those in carbonaceous chondrites: Clearly for these elements, carbonaceous chondrites provide a chemically unbiased sample of bulk solar system matter. Because the element silicon is the reference for comparison, it does not appear in the diagram.

Table 3-5. Abundances of metallic elements in chondritic meteorites:

	Percent of total metal atoms
Magnesium (Mg)	32
Silicon (Si)	33
Iron (Fe)	26
Aluminum (Al)	2.2
Calcium (Ca)	2.2
Nickel (Ni)	1.6
Sodium (Na)	1.3
Chromium (Cr)	0.40
Potassium (K)	0.25
Manganese (Mn)	0.20
Phosphorus (P)	0.19
Titanium (Ti)	0.12
Cobalt (Co)	0.10

In addition to telling us that the asteroids are made primarily of the four elements, the chemical composition of meteorites also tells us that the iron in these asteroids is not always mated with oxygen. In most meteorites, iron metal and iron sulfide also are found. As shown in Figure 3-3, despite the large changes in the ratio of the amount of iron in the metal (and sulfide) form to that in the oxide form from one chondrite type to another, the aggregate abundance of total iron remains nearly the same (that is, as the fraction of iron in metal form drops, the fraction of iron in oxide form rises correspondingly).*

Not all meteorites contain chondrules. The so-called achondrites are stony meteorites with textures much like igneous rocks found on Earth. Their appearance suggests that they are the products of a melting episode. In addition, metallic meteorites are found. When cut and polished, these objects show beautiful hexagonal patterns consisting of alternating bands of an iron alloy rich in nickel and an iron alloy poor in nickel (see Figure 3-4). Metallurgists recognize this pattern as one that forms when an alloy of iron and nickel is cooled very, very slowly. But they are unable to cool iron-nickel alloys anywhere near as slowly as is required to give these bold patterns. Thus, iron meteorites cannot be confused with man-made iron. Metallic iron is exceedingly rare in Earth surface rocks. As we shall see, the Earth's iron metal is stowed away in a central core.

In some cases a single meteorite has both metallic and stony parts. The stony parts have no chondrules. This evidence leads scientists to conclude that the achondrites and iron meteorites were formed through the melting of chondrites. Because liquid iron and liquid silicate are not miscible, were an asteroid made up of chondrites to have melted, the metal and silicate would have separated into an achondritic stony mantle and an iron core. Meteorites containing both appear to have been broken off from the mantle-core boundary of such an asteroid.

*There is an exception to this generalization. The enstatite chondrites have less iron than do the other four types shown in Figure 3-3.

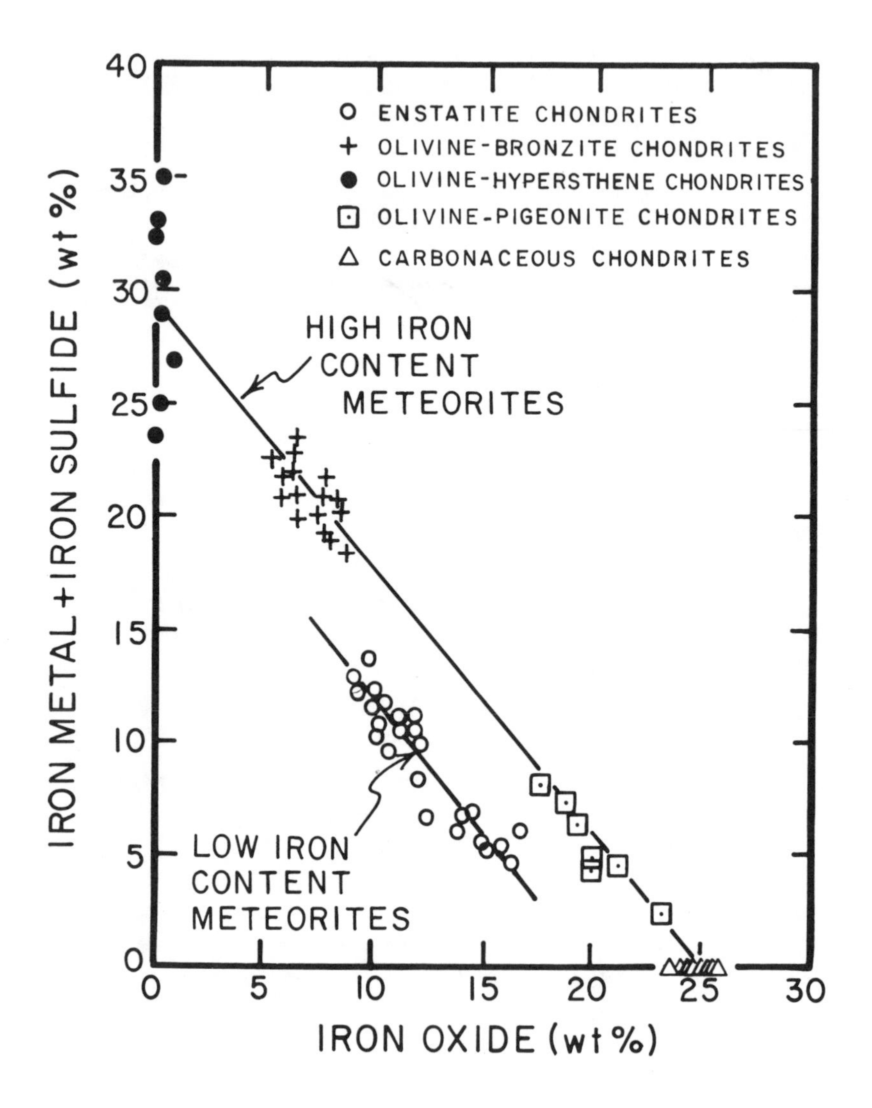

Figure 3-3. Gradations in chemical form of iron in chondrites: The range extends from all oxide in carbonaceous chondrites to all metal (or sulfide) in olivine-hypersthene chondrites. Except for enstatite chondrites, which are depleted in iron, the total iron content remains nearly the same.

Figure 3-4. Polished section of the Bagdad iron meteorite: The patterns made by adjacent nickel-rich alloy and nickel-poor alloy can be seen. These patterns, called Widmanstatten structures, are named after the man who pioneered their study. The maximum dimension is 10 centimeters.

Composition Differences Among the Terrestrial Planets and Meteorites

We now have the information needed to define the likely bulk chemical composition of the asteroids and terrestrial planets. Meteorites tell us that the ingredients of these objects must be stony matter made primarily of the elements silicon, magnesium, iron, and oxygen and metallic matter made primarily of iron. The bulk densities permit a rough estimate of the ratio of these two major ingredients in a given planet. Indeed, the rather large differences in these densities (see Figure 3-5 for summary) can be accounted for only by sizable planet-to-planet differences in the ratio of stone to metal. Mars appears to consist mainly of stony matter. Mercury appears to consist of about half iron and half stone. Earth and Venus have compositions falling midway between those of Mars and Mercury.

In the case of Earth, we have enough information to make a guess at an average chemical composition.* As shown in Table 3-6, the Earth has higher Fe/Si and Mg/Si ratios than do both high-iron and low-iron meteorites. We lack the information needed to assess the bulk chemical compositions for Mercury and Mars. Because of its great similarity to Earth, Venus likely has a chemical composition close to Earth's.

The Great Gas-Solid Separation

While it may seem strange that only four elements should dominate the terrestrial planets and asteroids, an examination of the solar abundances and chemical affinities for the elements shows that this is quite reasonable. The chemical differences between the minor planets and the Sun can only be accounted for through a separation between gases and solids. Elements with an affinity for gaseous form are thought to have been driven

*While no pristine samples of the Earth's interior are available for chemical analysis, clever detective work has allowed petrologists to make a case for the proportions of Fe, Si, Mg, and O for the Earth as a whole.

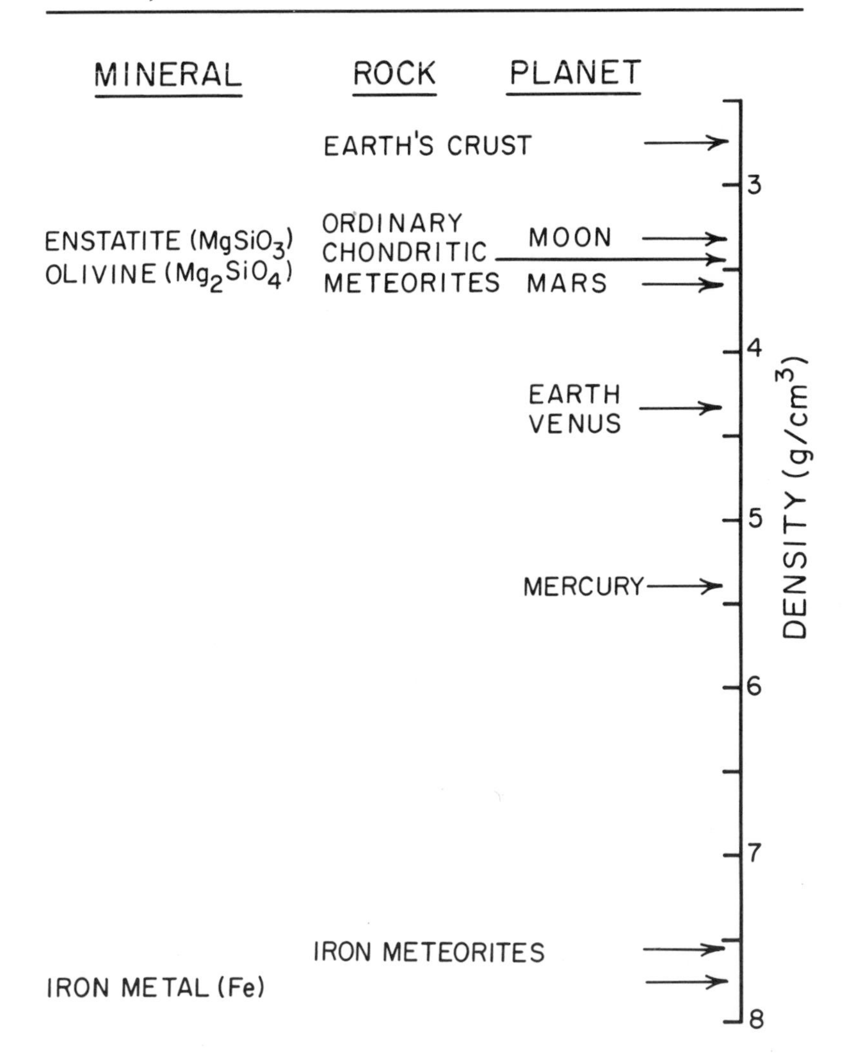

Figure 3-5. Bulk densities of various minerals, rocks, and planets: In the case of the planets, the densities shown have been corrected for gravitational compaction. The planet-to-planet density differences are in part the result of differences in the Fe/Mg+Si ratio and in part the result of differences in the iron/iron oxide ratio.

away from the solar nebula by the impacts of ions streaming out from the Sun. Elements in dust or ice particles were retained to form the planets. This separation is fundamental to planet formation. The stream of ions from the Sun may be likened to

Table 3-6. Comparison between the chemical compositions of various meteorite classes and that of the bulk Earth: Despite the great change in the fraction of iron in oxide form from class to class (see last column), the relative abundances of the three major constituent metals (i.e., Si, Mg, and Fe) remain nearly unchanged in the high-iron chondrites. As can be seen, the Earth is richer in magnesium and iron relative to silicon than are meteorites.

	Fraction of mass as SiO_2, MgO, FeO, and iron metal*	Relative atom abundance				Relative mass abundance				Fraction of iron in oxide form
		Si	Mg	Fe	O†	Si	Mg	Fe	O†	
LOW-IRON CHONDRITES										
Enstatite chondrites	.92	100	92	60	325	100	80	119	185	.55
HIGH-IRON CHONDRITES										
Carbonaceous chondrites	.78	100	104	84	380	100	90	167	216	.90
Olivine-Pigeonite chondrites	.92	100	101	78	357	100	87	155	203	.72
Olivine-bronzite chondrites	.91	100	96	79	340	100	83	157	194	.56
Olivine-Hypersthene chondrites	.92	100	84	82	285	100	73	163	162	.01
WHOLE EARTH										
Whole Earth	.94	100	131	126	359	100	114	250	199	.11

*FeS present in meteorites included with iron metal.

†Excludes oxygen associated with metals other than silicon, magnesium, and iron.

a strong wind blowing over an arid plain. Fine sediment is swept up and carried away by the wind. Pebbles are left behind because they are too heavy to be lifted by the passing air. Similarly, gas molecules were swept out of the disk of matter that orbited the Sun early in its history. The dust particles were too heavy to be moved by the impacting ions and remained behind.

Let us run through the list of elements in Table 3-7 and see why planets might be dominated by as few as four elements. The first element on the list is hydrogen. In the cloud of gas plus dust from which the planets formed, hydrogen atoms were present either as hydrogen gas (H_2) or as gases of carbon (CH_4), of nitrogen (NH_3), or of oxygen (H_2O). The Earth and its fellow terrestrial planets trapped only a tiny fraction of these gases; the rest was lost.

Helium exists only as a gas. Like the other noble gases, it seldom makes chemical unions with other elements. Hence virtually all the helium was lost. Even the very small amount of helium we do find today in the Earth's atmosphere and in gases escaping from the Earth's interior was not captured by the Earth; rather, it was produced within the Earth by the radioactive decay of the elements uranium and thorium.

The next three elements—lithium, beryllium, and boron—were produced in very small abundance by the synthesis mechanisms in stars. Their abundance relative to other solid-prone elements like magnesium, silicon, and iron is too small to permit them to be major constituents of the planets.

Carbon and nitrogen, in the presence of the large amounts of hydrogen gas in the planetary nebula, would have been in the form of methane (CH_4) and ammonia (NH_3). These gaseous compounds were largely lost.

While also attracted to chemical unions with hydrogen, the element oxygen is even more strongly attracted to chemical unions with the elements chemists refer to as metals. In the cloud from which the Sun and the planets formed there were five times as many oxygen atoms as all metal atoms taken together; hence, only about 20 percent of the available oxygen atoms were able to get the metal atoms that would be their first

Table 3-7. Relative abundance of the first 28 elements and their fates during the formation of the terrestrial planets:

Element Number	Element Name	Compound Solid	Compound Gas	Rel. Abundance In Sun*	Fate†	Rel. Abundance In Chondrites*
1	HYDROGEN		H_2	40,000,000,000	(1)	—
2	HELIUM		He	3,000,000,000	(1)	trace
3	LITHIUM	Li_2O		60	(3)	50
4	BERYLLIUM	BeO		1	(3)	1
5	BORON	B_2O_3		43	(2)	6
6	CARBON		CH_4	15,000,000	(1)	2,000
7	NITROGEN		NH_3	4,900,000	(1)	50,000
8	OXYGEN		H_2O**	18,000,000	(2)	3,700,000
9	FLUORINE		HF	2,800	(1)	700
10	NEON		Ne	7,600,000	(1)	trace
11	SODIUM	Na_2O		67,000	(2)	46,000
12	MAGNESIUM	MgO		1,200,000	(3)	940,000
13	ALUMINUM	Al_2O_3		100,000	(3)	60,000
14	SILICON	SiO_2		1,000,000	(3)	1,000,000
15	PHOSPHORUS	P_2O_5		15,000	(3)	13,000
16	SULFUR	FeS	H_2S	580,000	(2)	110,000
17	CHLORINE		HCl	8,900	(1)	700
18	ARGON		Ar	150,000	(1)	trace
19	POTASSIUM	K_2O		4,400	(2)	3,500
20	CALCIUM	CaO		73,000	(3)	49,000
21	SCANDIUM	Sc_2O_3		41	(3)	30
22	TITANIUM	TiO_2		3,200	(3)	2,600
23	VANADIUM	VO_2		310	(3)	200
24	CHROMIUM	CrO_2		15,000	(3)	13,000
25	MANGANESE	MnO		11,000	(3)	9,300
26	IRON	FeO,FeS,Fe		1,000,000	(3)	690,000
27	COBAL	CoO		2,700	(3)	2,200
28	NICKEL	NiO		58,000	(3)	49,000

*Relative to 1,000,000 silicon atoms.
†(1) Highly volatile; mainly lost;
(2) Moderately volatile; partly captured;
(3) Very low volatility; largely captured.
**Plus metal oxides.

choice as chemical mates. The remainder had to take their second chemical choice—hydrogen atoms. The gaseous water molecules thus formed were largely swept away. Only those oxygens with metal-atom mates were incorporated into solid phases.

After oxygen on the list in Table 3-7 come fluorine and neon. Fluorine atoms have a strong tendency to combine with hydrogen in the form of hydrofluoric acid (HF). Under the conditions that prevailed when the planets formed, HF was likely to be a gas. Like helium, neon shuns chemical unions. Hence, both elements were largely driven away.

So far we have gone through ten elements. Of these, six (hydrogen, helium, carbon, nitrogen, fluorine, and neon) formed gases and were largely lost. Three others (lithium, boron, and beryllium) had such small abundances as to be unimportant to the bulk composition of planetary material. Only oxygen was sufficiently abundant and sufficiently prone toward the formation of solid phases to become a major contributor to the terrestrial planets.

The next five elements on the list are all metals that prefer chemical unions with oxygen. Four of them (magnesium, aluminum, silicon, and phosphorus) were efficiently trapped in the solid material. The fifth (sodium) is moderately volatile; hence, some was lost. As can be seen in Table 3-8, silicon and magnesium both have isotopes that have nuclear-particle numbers divisible by four (^{24}Mg and ^{28}Si). These so-called alpha-particle nuclides were produced in stars in greater abundance than were neighboring nuclides. For this reason, magnesium and silicon are more abundant than sodium, aluminum, and phosphorus, which have no isotopes of the alpha-particle variety.

Next on the list is sulfur. Its situation is akin to that for oxygen. On one hand, it could form the hydrogen-bearing gas H_2S. On the other, it could combine with iron to form a solid, FeS. Our evidence from meteorites suggests that a significant fraction of the available sulfur was captured (as FeS).

The next two elements on the list, chlorine and argon, were largely lost as gases. Chlorine was in the form of hydrochloric acid (HCl), a gas. Argon, like helium and neon, is a noble gas which shuns chemical unions.

Next on the list are two more metallic elements, potassium and calcium. Calcium in the oxide form has a very low volatility. Like sodium, potassium is moderately volatile and hence was not captured with the same efficiency as were metals of low volatility. Despite its low abundance, potassium has an important role in Earth studies. In part because one of its isotopes (^{40}K) is radioactive and in part because it is a very important constituent of the Earth's crust.

So we see that in the second group of ten elements, five (magnesium, aluminum, silicon, phosphorus, and calcium) were largely captured. Three (sodium, potassium, and sulfur) were partly captured. Two (chlorine and argon) were largely lost.

Between calcium and iron there is a big sag in the abundance curve. Thus, although most of the elements in this interval are metals of low volatility, none is sufficiently abundant to challenge silicon or magnesium.

Table 3-8. Isotopic abundances of the big four planet-producing elements:* Note that for three of the elements the dominant isotope has a nucleus consisting of an integral number of α-particles (i.e., four in ^{16}O, six in $^{24}M_6$, seven in ^{28}Si). The terrestrial planets are 85 percent by weight ^{16}O, ^{24}Mg, ^{28}Si, and ^{56}Fe!

OXYGEN (8 protons)	^{16}O	^{17}O	^{18}O	
	99.76%	0.04%	0.20%	
MAGNESIUM (12 protons)	^{24}Mg	^{25}Mg	^{26}Mg	
	78.99%	10.00%	11.01%	
SILICON (14 protons)	^{28}Si	^{29}Si	^{30}Si	
	92.23%	4.67%	3.10%	
IRON (26 protons)	^{54}Fe	^{56}Fe	^{57}Fe	^{58}Fe
	5.8%	91.8%	2.1%	0.3%

*In reading the lines in solar rainbows we get estimates of the abundances of elements. Many elements have more than one isotope. Thus the solar evidence must be supplemented if we are to get the abundances of individual isotopes. This is done by measuring the ratios of the isotopes of a given element in meteorites or Earth surface materials. Although separations among the elements have caused large biases in the chemical composition of Earth surface rocks, these separations do not extend to the isotopes of a given element. The isotopes of an element are nearly chemically identical. Thus, the isotopic composition of a given element is generally found to be the same whether the sample analyzed is from the Earth or from a meteorite.

The abundance of iron, the ultimate product of nuclear fires, stands well above that of its neighboring elements. Also, none of its chemical forms in the early solar system was particularly volatile. As its abundance is similar to that of magnesium and silicon, it became one of the "big four" elements in the terrestrial planets.

Beyond iron, the abundance of the elements drops rapidly with increasing proton number. Only nickel is sufficiently abundant to be important. As shown in Table 3-5, nickel along with aluminum, calcium, and sodium make up the second abundance group.

Thus we see that a combination of nuclear physics (which sets the relative abundances of the elements) and inorganic chemistry (which sets the chemical form of the elements in the planetary nebula) dictated that rocky planets like Earth consist primarily of the elements oxygen, magnesium, silicon, and iron. One might then ask why the ratio of Mg to Si to Fe is not identical in the terrestrial planets. The answer must be that at some stage in the planet-formation process, the material must have been so hot that even iron, magnesium, and silicon were at least in part in volatile form and hence lost from the solar nebula along with the other gases. This partial loss separated these elements from one another. The chemical differences among the planets tell us that the nature and extent of this separation must have varied from place to place in the solar nebula.

K-to-U Ratios: A Measure of Volatile Retention

The differences in density among the terrestrial planets dictate that the situation was more complicated than a clear-cut separation between a set of elements forming gaseous compounds and another set of elements forming dust particles. As we shall see in Chapter Seven, the terrestrial planets captured traces of the elements whose major chemical form was gaseous. Variations in the temperature at any one place in the nebula with time, and at any one time with distance from the Sun, allow for differences in the efficiency with which any given element was captured. A particularly useful index of this capture efficiency is provided by the ratio of a planet's potassium (K) to uranium (U) content. These two elements differ in capture efficiency because potassium compounds are more volatile than uranium compounds. However, these two elements showed a remarkable coherence in their chemical pathways during the melting episodes that caused the planets to become chemically layered. We know this because a wide variety of volcanic rocks from the Earth's surface have nearly the same K-to-U ratio. Thus, despite our inability to directly sample the Earth's interior we feel confident that the K-to-U ratio for the Earth's bulk is similar to that observed in Earth's surface rocks. As shown in Table 3-9, the Earth has a K/U ratio seven times lower than that for chondrule-bearing meteorites.

Both K and U have long-lived radioisotopes which emit gamma rays. Since these powerful electromagnetic radiations are able to escape the rocks in which the K and U are contained, they can be detected by instruments dropped to the planetary surface. This is important because it permits the chemical analyses of planets from which we cannot recover samples. Unmanned spacecraft landed on Mars and on Venus have telemetered back gamma-ray-based estimates of the K-to-U ratio in the surface rocks of these planets. As for the Earth, these ratios are taken to be indicative of those for the bulk planets. As can be seen in Table 3-9, large differences are found. Mars has a K-to-U ratio about 20 times lower than that for chondrites. By contrast, Venus has a K/U ratio only about six times lower than that for chondrites. Thus, the materials making up Mars appear to have been heated with the most intensity. Earth and Venus materials experienced heating episodes of intermediate intensity. The material in chondritic meteorites experienced the lowest intensity of heating.

The potassium content of ordinary chondrites is only slightly different from that of carbonaceous chondrites. However, for elements more volatile than potassium, the difference in concentration between ordinary chondrite and carbonaceous chondrite becomes quite large (see Figure 3-6). Thus, we should expect even larger planet-to-planet differences for elements more volatile than potassium. Herein lies one of the most important factors influencing the potential habitability of a planet. As we

Table 3-9. Potassium-uranium ratios for four regions of the solar system: As can be seen, the extent to which moderately volatile potassium was separated from low volatility uranium varied considerably from place to place in the solar nebula.

Planet	K to U Ratio
Chondritic meteorites	70,000
Earth	10,000
Venus	≃ 12,000
Mars	≃ 3,000

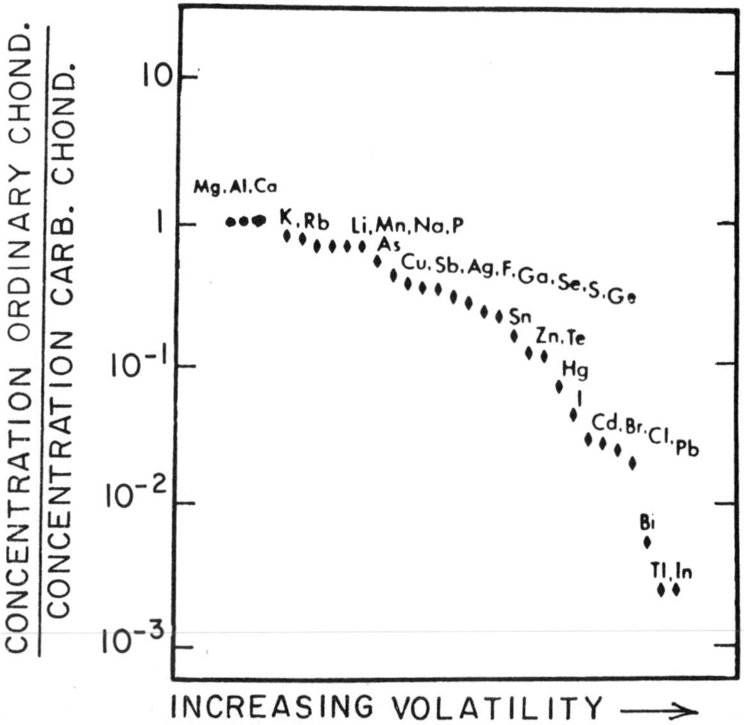

Figure 3-6. Depletion of volatile elements in ordinary chondrites: For each element the ratio of its concentration in ordinary chondrites to that in carbonaceous chondrites is shown. While the exact order of the elements with regard to volatility is subject to interpretation to a large extent, the greater the volatility of an element, the greater the degree to which it was lost during the baking process.

shall see, life of any kind, and in particular life capable of generating a civilization, requires that certain of these volatile elements be present on the planetary surface in just the right amount and form.

Table 3-10. Densities and melting temperature of possible planet-forming solids:

Compound	Number of nuclear particles per atom	Density of solid gm/cm³	Melting point of solid °C
		ICES	
CH_4	3.2	0.4	−184
NH_3	4.2	0.7	−78
H_2O	6.0	1.0	0
		OXIDES	
SiO_2	20	2.7	1710
Mg_2SiO_4	20	3.2	1200
		METAL	
Fe	56	7.9	1540

The Major Planets

As mentioned above, the major planets have bulk densities which, even without correction for gravitational squeezing, are far smaller than those for the terrestrial planets. The densities of the major planets can only be accounted for if they consist to a large extent of hydrogen. One way to get hydrogen into a planet is to have the conditions cold enough to permit the gases CH_4, NH_3, H_2O, and H_2S to freeze to ices. The temperatures required are listed in Table 3-10.

Although helium and hydrogen liquefy only at temperatures very close to absolute zero, it is still possible to get these gases into planets. If a planet grows big enough while its surroundings are still rich in hydrogen and helium gas, it can pull in these gases by sheer gravitation (just as the Sun did). Indeed, the densities of Jupiter and Saturn are so low that they can only be explained if the planets consist mainly of hydrogen and helium gas. Thus, these two planets must have competed with the Sun for the major gaseous ingredients of the source cloud. Jupiter was sufficiently successful in this regard that it came close to achieving a size large enough to have its own nuclear furnace. Had Jupiter become a bit larger, our Sun would be a binary star (a pair of stars orbiting one another). Many such stellar pairs are seen in our galaxy.

Summary

The organization of the planets tells us that they must have formed from a disk of matter that circled the early Sun. Their chemical compositions (see Table 3-11 for summary) tell us that this disk was sufficiently hot in its inner regions that only elements forming compounds that volatilize above 1000 or so degrees centigrade accumulated into solid bodies, and that it was sufficiently cold in the disk's outer regions to permit compounds that freeze only below 0° C to accumulate. The slow spin rate of the Sun also tells us that the matter built into the planets somehow managed to "capture" almost all of the system's angular momentum. While informative, this information is not sufficient to permit a choice among the many possible scenarios that have been put forward regarding the sequence of events that transformed the initial gas cloud into a star with nine planets.

The best solution would be to study the planetary systems of a host of newly born stars elsewhere in the galaxy. But we are unable to see planets being formed; indeed, so far we can't see planets at all outside our own solar system. There is as yet no confirmed sighting of a planet on any neighboring star. While our logic tells us that they must be there, they hide in the black abyss of space. The hope is that new sensors mounted on orbiting satellites will unveil this unseen world to us.

Table 3-11. Approximate compositions of the objects in the solar system: Note that the Sun's mass is nearly 770 times that of the combined planets. The environment in the region of the major planets was sufficiently cold when they formed so that they accumulated ices as well as silicate and iron. These four planets became sufficiently massive to pull in the noncondensible gases H_2 and He.

Object	Total Matter	Metals† Fe, Ni, . . .		Oxides† SiO_2, MgO, FeO, . . .		Ices† H_2O, CH_4, NH_4, H_2S, . . .		Gases H_2 + He	
	Mass 10^{27}gm	%	Mass 10^{27}gm	%	Mass 10^{27}gm	%	Mass 10^{24}gm	%	10^{27}gm
Sun	1,990,000	0.1	—	0.2	—	1.2	—	98.5	—
Mercury	0.33	50	0.16	50	0.17	—	—	—	—
Venus	4.87	30	1.46	69	3.36	≈1*	≈0.05*	—	—
Earth	5.97	29	1.73	69	4.12	≈2*	≈0.12*	—	—
Mars	0.64	10	0.06	90	—	—	—	—	—
Asteroids	0.0002	15	0.00003	85	0.00017	—	—	—	—
Jupiter	1900	≈4	≈80	≈9	≈170	≈5	≈100	≈82	≈1550
Saturn	570	≈7	≈40	≈14	≈80	≈12	≈70	≈67	≈380
Uranus	88	≈8	≈7	≈17	≈15	≈60	≈53	≈15	≈13
Neptune	103	≈6	≈6	≈14	≈14	≈70	≈73	≈10	≈10

†Likely as solid forms when accumulated by the planets. In the Sun the temperatures are so high that all elements are in gaseous form.

*Likely to have accumulated in some non-ice form.

Supplementary Readings

The Solar System, *by John Wood, 1979, Prentice-Hall, Inc.*

A book covering most aspects of the origin of the Sun and its planets.

Meteorites, *by John Wasson, 1974, Springer-Verlag.*

A book covering observation on meteorites and theories of their origin.

Chondrules and Their Origins, *edited by Elbert King, 1983, Lunar and Planetary Institute Publication.*

A series of papers concerning chondrules and their origins written for specialists.

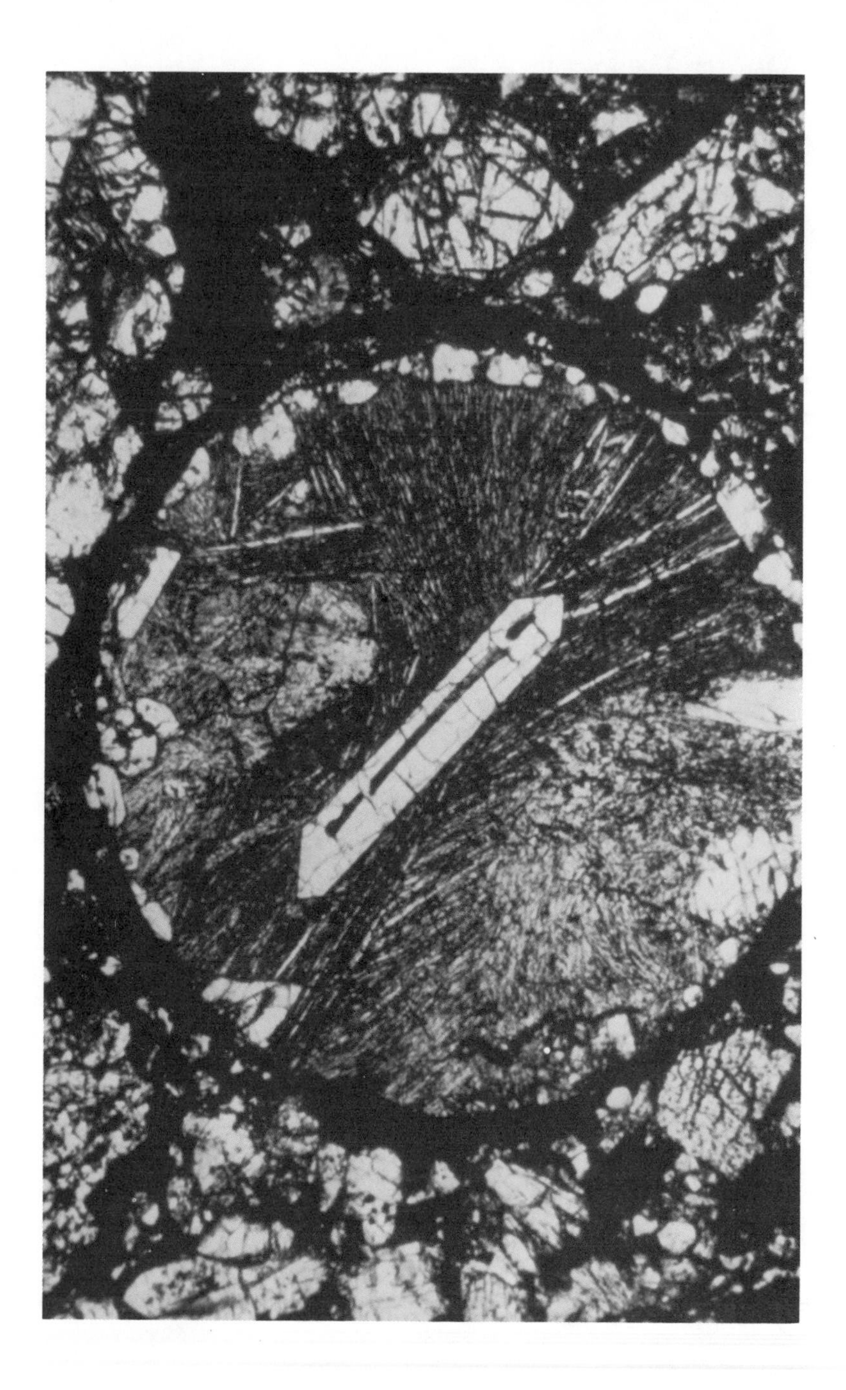

Chondrule

Chapter Four

The Schedule:

The Timing of Raw-Material Production and Heavy Construction

Some of the atoms ejected from stellar interiors into the surrounding galaxy by supernovae are radioactive. These atoms are of special interest to the historians of our solar system because they provide a means for telling time.

By comparing the present-day abundance ratios for certain long-lived radioisotope pairs with the abundance ratios computed for production in red giants, it is possible to get some sense of when the elements making up our planetary system were synthesized. If the frequency of supernovae is assumed to have remained more or less constant throughout the history of our galaxy, then these isotope pairs tell us that the galaxy must have formed about 14 billion years ago. This result fits comfortably with the age for the universe based on the the red shift.

Because they contain long-lived radioisotopes, all rocks formed from molten liquids are clocks. By measuring the concentrations of the radioactive isotope and of its decay product, it is possible to establish precisely the time when the rock crystallized. Because of the chondrules they contain, certain of the stony meteorites are thought to record the time the solar system formed. They read 4.6 billion years.

Differences in the abundances of isotopes formed by the decay of now-extinct radioisotopes among various solar-system materials tell scientists that element formation took place in the vicinity of the gas cloud from which our Sun formed just prior to its formation. This finding led to the speculation that the collapse of this long-quiescent cloud was triggered by the shock wave sent out by the supernova.

Introduction

In contrast to the sparsity of our knowledge regarding how planets formed, we do have a very precise estimate of when they formed and an upper limit on how long it took to complete this process. This information is derived from the abundances of long-lived radioisotopes and their daughter products in meteorites. The gradual transformation from radioactive parent isotope to stable daughter isotope is somewhat akin to the flow of sand through an hourglass. After considerable effort, scientists working in the field of geochemistry have learned to read these natural hourglasses accurately. From these readings we know that the material in meteorites accumulated 4.56 billion years ago.

Before we probe the methods used to obtain this information let us see how it fits into the context of the evolution of the universe. As discussed in Chapter One, the red shift-versus-distance relationship tells us that the big bang occurred about 15 billion years ago. Thus, something like 10 billion years elapsed before the cloud of dust and gas destined to produce our Sun and its planets began its collapse. This provides plenty of time for the formation of our galaxy and for the many generations of red-giant stars required to build the suite of elements we see in our Sun.

Listed in Table 4-1 are the radioactive nuclides found in nature that have half-lives of 0.1 billion to 1000 billion years. These isotopes were produced in stars by neutron irradiation. They are destined to transform into a stable neutron-proton combination. Because of their very long half-lives, however, they persist many billion years after being blown into the galaxy from their parent stars. These isotopes serve as our clocks.

The Birth Date of Meteorites

Our knowledge of the age of the solar system comes from determining the time of crystallization of the minerals that constitute chondritic meteorites. As discussed in the previous chapter, these objects formed during the early stages of planet production. All the materials available to us from the Earth have been remelted and recrystallized one or more times since the planets formed. During each of these meltings their radio clocks were reset. Hence, they no longer carry a record of the age of

Table 1. Radionuclides of stellar origin found in meteorites.*

Radionuclide	Half-life (Billions of years)	Stable daughter product
^{40}K	1.28	^{40}Ca and ^{40}Ar
^{87}Rb	49	^{87}Sr
^{138}La	110	^{138}Ce and ^{138}Ba
^{147}Sm	110	^{143}Nd
^{176}Lu	29	^{176}Hf
^{187}Re	50	^{187}Os
^{232}Th	14	^{208}Pb
^{235}U	0.72	^{207}Pb
^{238}U	4.47	^{206}Pb

*A number of other radioisotopes exist but their half-lives are too long to be of use in geochronology. For all practical purposes they can be considered stable.

the solar system. The presence of chondrules provides evidence that chondritic meteorites never melted; hence, their clocks record the age of the solar system.

The most precise estimate for the age of meteorites comes from measurements of the abundances of radioactive rubidium (^{87}Rb) and of its strontium daughter product (^{87}Sr). The ^{87}Rb atoms are equivalent to the sand remaining at the top of an hourglass and the ^{87}Sr atoms to the sand accumulated at the bottom of the glass. Were the situation exactly analogous to that in an hourglass, it would be easy to read the age. The user makes sure that there is no sand at the bottom of his hourglass at time zero. It is all at the top. Were we to come upon someone else's glass when part, but not all, of the sand had drained down, we could estimate when its missing owner had started it by measuring both the amount of sand in the bottom of the glass and the rate at which grains were falling through the neck of the glass. Seemingly we could do the same thing with meteorites. By measuring the rate at which ^{87}Rb atoms are decaying in a meteorite sample and the number of ^{87}Sr atoms, we could get the time elapsed since the meteorite formed. Unfortunately, meteorites are not ideal hourglasses. Only part of the ^{87}Sr found in a meteorite was generated by the decay of ^{87}Rb within the meteorite. The rest was present in the meteorite when it formed. This ^{87}Sr was produced either directly in stars or by the decay of ^{87}Rb between the time of nucleosynthesis and the time the meteorite formed. To read the hourglass properly it is necessary to distinguish between the ^{87}Sr present in the meteorite when it formed (the so-called common ^{87}Sr) and that generated within the meteorite by the decay of ^{87}Rb (the so-called radiogenic ^{87}Sr).

This is not an easy task. To make this point, the relative abundances of atoms of the isotopes of the elements rubidium and strontium in bulk chondritic meteorite samples are shown in Figure 4-1. For convenience, the abundance of each is given relative to 1000 ^{86}Sr atoms (that is, for each 1000 ^{86}Sr atoms in the meteorite there are 8370 ^{88}Sr atoms and 820 ^{87}Rb atoms).

Now, in the past there must have been more ^{87}Rb atoms (they gradually decay away) and correspondingly fewer ^{87}Sr atoms in the meteorite. To see the magnitude of this decay effect, we have calculated the isotopic composition of the meteorite as it was 1, 2, 3, 4, and 5 billion years ago, as shown in Figure 4-1. Because ^{84}Sr, ^{86}Sr, ^{88}Sr, and ^{85}Rb are neither radioactive nor the daughters of a radioisotope, their abundances remain exactly the same. For each billion years we go back into the past, the abundance (relative to 1000 ^{86}Sr atoms) of ^{87}Rb goes up by 12 atoms and that of ^{87}Sr goes down by 12 atoms. Had these meteorites been in existence 5 billion years ago they would have had $(820 + 5 \times 12)$ or 880 ^{87}Rb atoms and $(754 - 5 \times 12)$ or 694 ^{87}Sr atoms per 1C00 ^{86}Sr atoms. The point to be made is that even in a 5-billion-year period only about 8 percent of the ^{87}Sr atoms found in chondritic meteorites could have been produced by the decay of ^{87}Rb. Thus, most of the ^{87}Sr in meteorites is "common." This indicates that the job of separating the two types is an extremely important one. Most of the sand was already at the bottom of the glass at time zero! Fortunately, there is a way to do this job very accurately.

The key to the separation of the contributions of the common and the radiogenic ^{87}Sr lies in the observation that the various mineral types that make up meteorites contain different concentrations of the elements rubidium and strontium. Both these

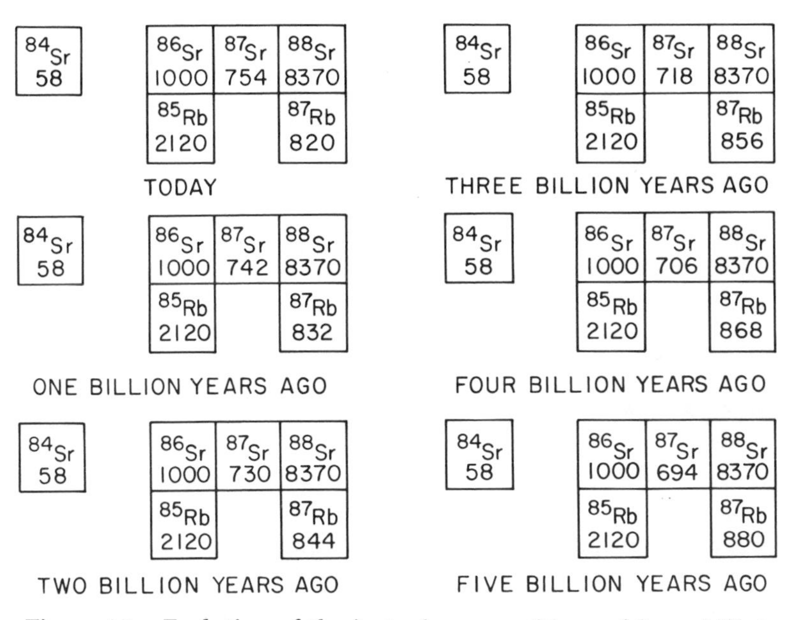

Figure 4-1. Evolution of the isotopic compositions of Sr and Rb in chondrites: The boxes in each of the six groupings on this diagram represent the naturally existing isotopes of elements strontium and rubidium. They are arranged as they would be on a chart of the nuclides. In the upper left grouping are shown the relative (to 1000 atoms of ^{86}Sr) amounts of these nuclides found today in a typical chondritic meteorite. Using the half-life of ^{87}Rb (49 billion years), it is possible to calculate the amount of ^{87}Rb in this chondritic meteorite 1, 2, 3, 4, and 5 billion years ago. Since one ^{87}Sr atom was produced for each ^{87}Rb atom that decayed, it is also possible to calculate how many ^{87}Sr atoms were present at each of these times. Since the remaining four isotopes are not radioactive and are not produced by the radiodecay of any other nuclide, their amounts do not change with time. The purpose of this exercise is to show that if the age of meteorites is in the range from 1 billion years to even 10 billion years, no more than about 15 percent of their ^{87}Sr atoms can have come from radiodecay. Most of the ^{87}Sr must have been there when the meteorite formed!

elements are so low in concentration in meteorite material (Sr averages 10 parts per million by weight and rubidium 3 parts per million by weight) that they do not form minerals of their own. Rather, they appear as trace contaminants in minerals made primarily of other elements. Just as hermit crabs hide out in suitable abandoned shells, strontium and rubidium atoms hide out in minerals designed to house the more abundant elements. Rubidium and strontium differ from one another in their chemical affinities; hence their criteria for selecting homes are not the same. Because of this, some minerals ended up with Rb/Sr ratios higher than that for the bulk meteorite material and others ended up with Rb/Sr ratios lower than that for the bulk meteorite. This is the key.

Ideally, one would hope to find one mineral in which no rubidium whatsoever existed. In such a mineral all the ^{87}Sr would have to be common. Its ^{87}Sr abundance (relative to 1000 ^{86}Sr atoms) could then be used as a reference to correct for common ^{87}Sr in other minerals which did contain rubidium. As it turns out, no such mineral has been found.

Fortunately there is another way to go about finding the isotopic composition of the strontium in this hypothetical rubidium-free mineral. This is accomplished by making isotopic analyses on a series of mineral types hand-picked from a crushed piece of meteorite. In Figure 4-2 a graph of the ^{87}Sr content against the ^{87}Rb content (both relative to 1000 ^{86}Sr atoms) is shown. Each point on this graph represents the results of measurements on a mineral separated from a chondritic meteorite. As would be expected, the higher the ^{87}Rb abundance in the mineral, the higher the ^{87}Sr abundance. In fact, all the measurement points fall very nicely on a straight line, demonstrating a perfect proportionality. If this line is extended until it reaches the left-hand margin of the diagram (that is, ^{87}Rb abundance equal to zero), we have the sought-after primordial isotopic composition of strontium. When the solar system formed, there were 700 atoms of ^{87}Sr for each 1000 atoms of ^{86}Sr. It does not take much of an act of faith to believe that this is the ^{87}Sr content that would be found in a mineral free of the element rubidium! As the isotopes of a given element are almost identical chemically, they were not separated one from the other during the formation of meteorites. This lack of separation is verified by the constancy of the ratio of ^{84}Sr to ^{86}Sr in all planetary samples examined to date. Thus, the strontium in all the minerals must have had the same isotopic composition when meteorites formed (700 ^{87}Sr atoms per 1000 ^{86}Sr atoms).

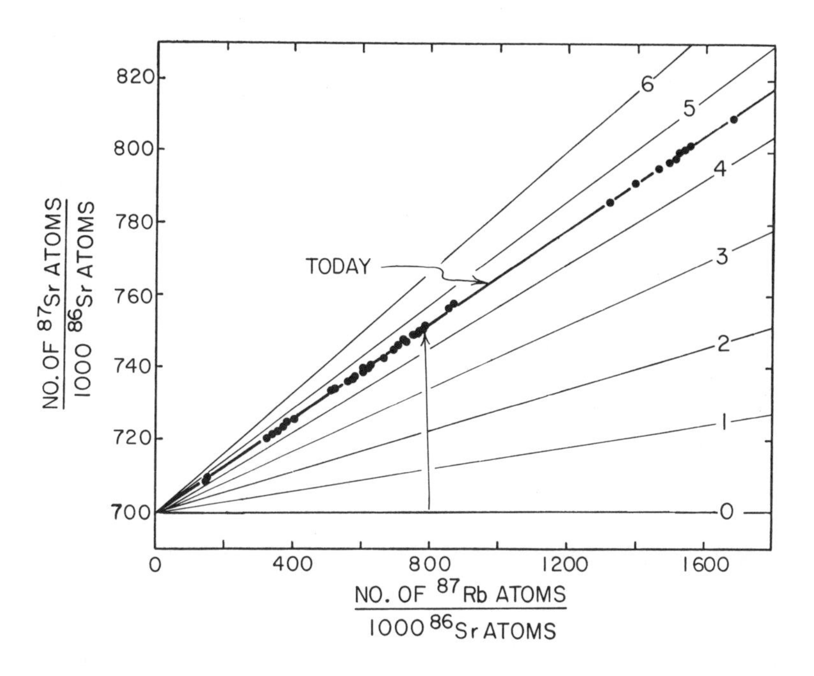

Figure 4-2. Evolution of strontium isotope composition in minerals of differing rubidium contents: The light lines on this diagram show the evolution with time (billions of years) of the ^{87}Sr and ^{87}Rb in meteorites. The measurements on mineral grains separated from chondritic meteorites tell us two things. First, they tell us that there were 700 ^{87}Sr atoms for each 1000 ^{86}Sr atoms in the strontium present in the solar nebula. Second, they tell us that these meteorites formed very close to 4.56 billion years ago. The former is derived from the intercept of the straight line that passes through the measured values. The latter is derived from the slope of the line passing through these points. Each grain followed a time trend parallel to that for the arrow shown on the diagram. At the time the solar system formed all the grains had compositions falling along the line marked zero, i.e., they had a range of ^{87}Rb to ^{86}Sr ratios, but all had 700 ^{87}Sr atoms per 1000 ^{86}Sr atoms. With time each grain increased in ^{87}Sr content (and decreased in ^{87}Rb content). This increase was in proportion to its ^{87}Rb content.

If we go back to Figure 4-1, we see that our hypothetical carbonaceous chondrite would have had 706 atoms of ^{87}Sr for each 1000 ^{86}Sr atoms 4 billion years ago, and 694 ^{87}Sr atoms for each 1000 ^{86}Sr atoms 5 billion years ago. If it started with 700 ^{87}Sr atoms per 1000 ^{86}Sr atoms, then the minerals in this meteorite must have formed about 4.6 billion years ago.

The arrow in Figure 4-2 shows the pathway followed by the isotope ratios in one of the meteorite samples as it aged. When the mineral designated by the arrow formed it had a ^{87}Rb abundance of 800 and a ^{87}Sr abundance of 700. Today it has a ^{87}Rb abundance of 746 and a ^{87}Sr abundance of 754. In another 4.6 billion years it will have a ^{87}Rb abundance of 692 and ^{87}Sr abundance of 892 (relative to 1000 ^{86}Sr atoms).

In Figure 4-3, the equations are given which allow the age of the meteorites to be calculated from the slope of the line drawn through the measurement points. The result is 4.56 $\pm$ 0.01 billion years! Similar analyses have been made on a variety of meteorites. As summarized in Figure 4-4, within the uncertainty in the measurements, all the meteorites give the same age. Not only does this fortify our confidence in the validity of the estimate of the age of these objects, it also demonstrates that all meteorites formed at very nearly the same time.

A very analogous procedure can be followed to obtain an independent estimate of the age of meteorites using the ^{147}Sm-^{143}Nd nuclide pair. The radionuclide ^{147}Sm has a half-life of 110 billion years. It produces ^{143}Nd by α-decay. Analyses of the abundances of the samarium and of the neodymium isotopes are made on minerals separated from meteorites. When the ratio of ^{143}Nd to ^{145}Nd abundance is plotted against the ratio of ^{147}Sm to ^{145}Nd abundance, again the points fall on a straight line. The intercept of this line gives the isotopic composition of the element neodymium at the time the solar system formed. The slope of the line drawn through the points gives the age of the meteorite. Again the answer comes out to be 4.6 billion years.

Age of the Elements

The relative abundances of the long-lived radioisotopes in meteorites also have something to tell us about when the heavy elements heavier than hydrogen and helium were produced. As we have no solid objects to act as hourglasses to record the age of these events, we turn to the abundance ratios of pairs of long-lived radioisotopes. The idea is that as the two isotopes have different half-lives, their abundance ratio will change with time.

Figure 4-3. Derivation of equation used to calculate the age of meteorites:

The concentrations of the isotopes are related as follows:

$$^{87}Sr\Big)_{today} = {}^{87}Sr\Big)_{begin} + \left[{}^{87}Rb\Big)_{begin} - {}^{87}Rb\Big)_{today}\right]$$

$$^{87}Rb\Big)_{begin} = {}^{87}Rb\Big)_{today}\, e^{\frac{t}{\tau}}$$

where t is the age of meteorite, τ is the mean life of ^{87}Rb atoms (i.e., $t_{1/2}/.693$ or 70×10^9 years). Thus:

$$^{87}Sr\Big)_{today} = {}^{87}Sr\Big)_{begin} + {}^{87}Rb\Big)_{today}\left(e^{\frac{t}{\tau}} - 1\right)$$

If t is much less than τ, then the following approximation can be made

$$e^{\frac{t}{\tau}} - 1 \simeq \frac{t}{\tau}$$

Hence:

$$^{87}Sr\Big)_{today} = {}^{87}Sr\Big)_{begin} + \frac{t}{\tau}\, {}^{87}Rb\Big)_{today}$$

$$\frac{^{87}Sr}{^{86}Sr}\Bigg)_{today} = \frac{^{87}Sr}{^{86}Sr}\Bigg)_{begin} + \frac{t}{\tau}\,\frac{^{87}Rb}{^{86}Sr}\Bigg)_{today}$$

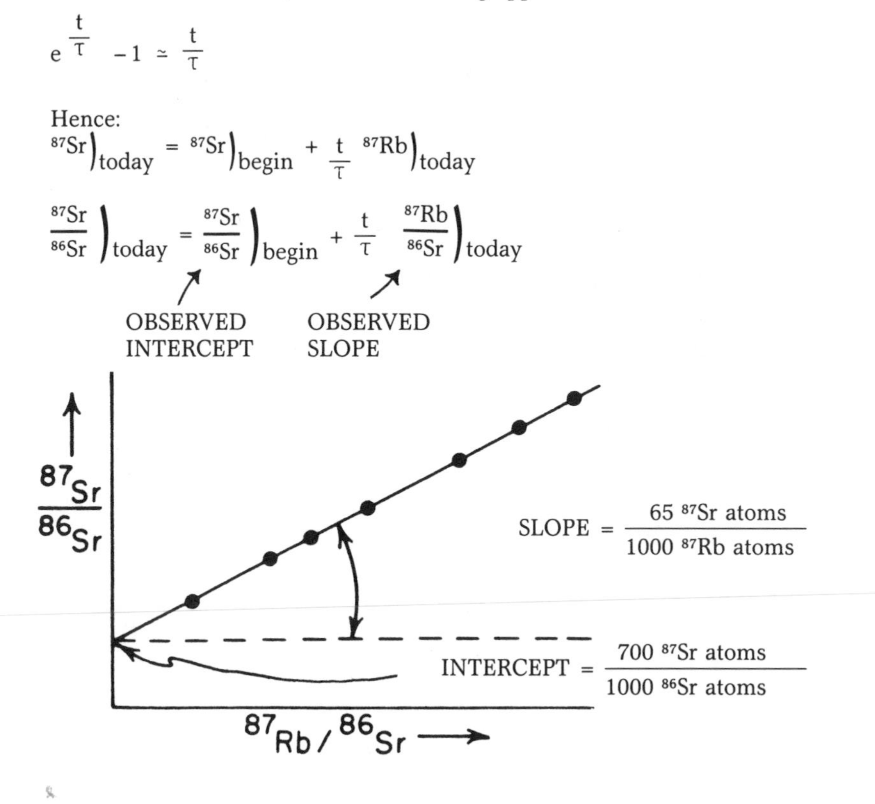

The meteorite measurements yield a slope of 0.065. Hence, $t = 0.065\tau = 0.065 \times (70 \times 10^9) = 4.55 \times 10^9$ years.

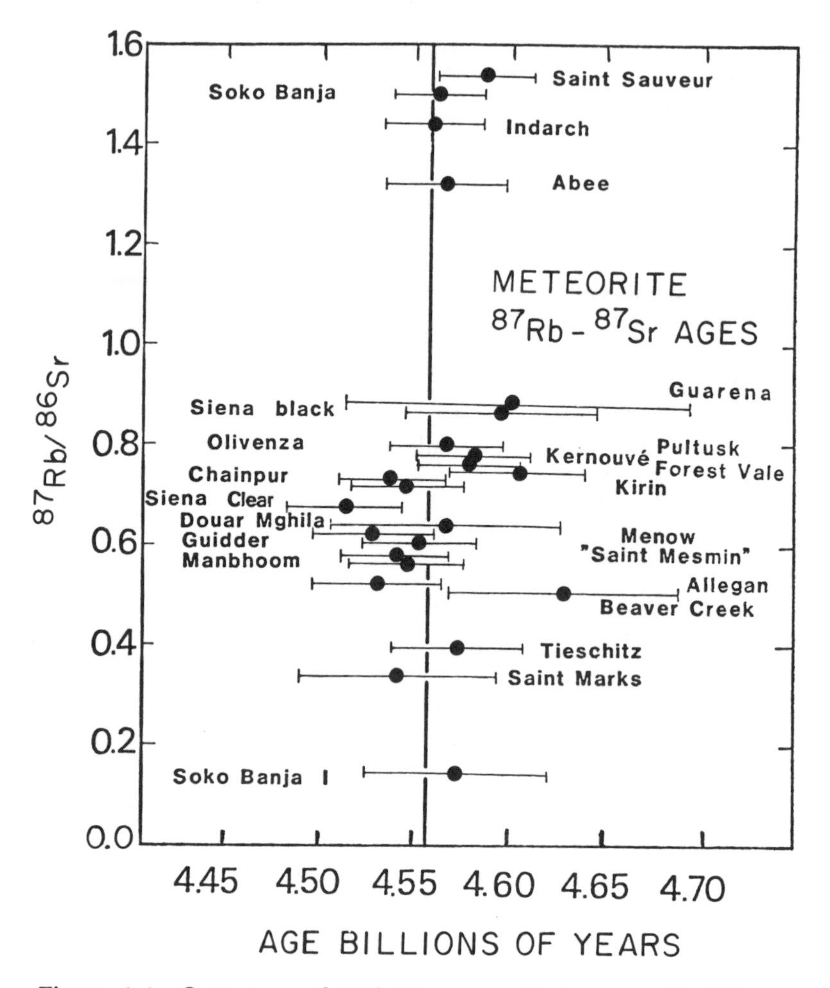

Figure 4-4. Summary of ^{87}Rb-^{87}Sr ages obtained on 19 different meteorites: As can be seen, all the results lie between 4.52 billion and 4.63 billion years. The mean of all the measurements is 4.56 billion years (shown by vertical black line). As the uncertainty in each measurement (shown by horizontal bars) in all but three cases spans this mean, there is no evidence for significant differences in the age of these objects.

In order for this approach to work, the isotope pair must satisfy three criteria:

1) The nuclides must have half-lives in the range of 0.2 billion to 20 billion years. If the half-life falls below this range,

then the nuclide will no longer be detectable. If it falls above this range, then too little change in its abundance will have occurred.

2) The two nuclides must not have been separated significantly during the formation of the planets.
3) Since the age of the heavy elements is obtained by comparing the production ratio of two such isotopes by the r-process with today's observed ratio, an accurate estimate of the relative abundances of the nuclides of interest in the debris leaving a supernova explosion must be obtainable on theoretical grounds.

These restrictions lead us to three of the nuclides listed in Table 4-1: ^{235}U, ^{238}U, and ^{232}Th. All three have half-lives in the right range. Since two of the nuclides are isotopes of the same element, they were not separated from one another during the formation of the planets. The third is an isotope of thorium. Like those of uranium, thorium compounds have very low volatilities. Thus, it is unlikely that thorium and uranium were separated from one another during the accumulation of chondrites.

Physicists who have tried to duplicate the r-process buildup in their computers calculate that these three isotopes should be produced in roughly the following proportions:*

$$^{235}U : {}^{238}U : {}^{232}Th$$
$$790 : 525 : 1000$$

By contrast, we find these isotopes today in carbonaceous chondrites in the following proportions:

$$^{235}U : {}^{238}U : {}^{232}Th$$
$$1.9 : 265 : 1000$$

There are now about four hundred times fewer ^{235}U atoms and about two times fewer ^{238}U atoms per 1000 ^{232}Th atoms than expected for fresh supernovae debris. The difference is in the expected direction: ^{235}U (half-life 0.7×10^9 years) has decreased the most relative to ^{232}Th; ^{238}U (half-life 4.5×10^9 years) has

*In these calculations we use 1000 atoms of ^{232}Th (half-life 14 billion years) as the reference abundance.

decreased by a much smaller amount relative to ^{232}Th. Clearly, a long time has passed since the solar system's uranium and thorium were formed.

Having established that the solar system is 4.6×10^9 years old we can calculate the relative abundances of these three nuclides at the time meteorites formed. We must use radiodecay equation to correct for the loss during this time. The equation is

$$\frac{N^{then}}{N^{now}} = e^{\frac{\Delta t}{\tau}}$$

where Δt is the time elapsed since the formation of the solar system (4.6×10^9 years) and τ is the mean life of the isotope of interest. The answers we get are as follows: at the time the solar system formed, ^{235}U was 80 times more abundant than it is today, ^{238}U was 2.0 times more abundant, and ^{232}Th was 1.25 times more abundant. Thus the relative abundances of these three nuclides at the time the solar system formed must have been

^{235}U	:	^{238}U	:	^{232}Th
80×1.9	:	2×265	:	1.25×1000

which becomes

^{235}U	:	^{238}U	:	^{232}Th
152	:	530	:	1250

or in terms of 1000 ^{232}Th atoms,

^{235}U	:	^{238}U	:	^{232}Th
122	:	424	:	1000

These abundances can be compared with the abundances calculated for fresh supernova debris:

^{235}U	:	^{238}U	:	^{232}Th
790	:	525	:	1000

While closer, the abundances at the time the earth formed still show substantial evidence for radiodecay. This difference indicates that a substantial amount of time must have elapsed between the origin of the elements and the formation of our Sun.

We can get a rough idea of the magnitude of this time period by considering the uranium isotopes. To go from the $^{235}U/^{238}U$ uranium-production ratio (790/525) to the ratio at the time the solar system formed (122/424) requires that the uranium have aged for 2.1 billion years. Thus, had all the uranium in our solar system formed during a single supernova explosion, then the time of this event must have been about 6.7 billion years ago (4.6 + 2.1).

However, as this calculation assumes that the elements heavier than the hydrogen and helium formed during a single event, it cannot be correct. We know that these heavy elements must have been produced by a long series of events stretching from close to the time our galaxy formed until close to the time the Sun formed. Since supernovae are observed to occur about once each hundred years in our galaxy, it is likely that about 10^8 such events occurred during this interval (10×10^9 yrs divided by 100 yrs). Unfortunately, we do not know how the frequency of supernova explosions has changed over the history of our galaxy. Faced with this uncertainty, scientists fall back on a hunch. Their hunch is that red giants formed and died with about the same frequency over the galaxy's entire history. If so, then the rate of element production has remained constant since our galaxy formed.

In this case, the amount of each of the stable nuclides (heavier than H and He) would have increased steadily with time. However, in the case of the radioisotopes of interest, as the galaxy aged, this climb in abundance would begin to level off and eventually the abundance would become constant. The reason is that eventually the loss of the isotope via radioactive decay would just balance the new production of the isotope in stars.

An analogy will aid in understanding this concept. When the Tutankhamen exhibition was shown in the Metropolitan Museum of Art in New York City, viewers were admitted to the gallery at a controlled rate. Just after opening each day the gallery would be nearly empty. As time passed, the number of viewers would steadily increase. However, as the average viewer would spend only one hour at the exhibit, this buildup would level off and by early afternoon the number of people leaving the exhibition would just balance the number of people entering.

Now let us complicate the analogy a bit in order to show why the ratio of the abundances of two long-lived radioisotopes could be used to tell us when production of heavy elements began. Let's assume that half of the admittees were tourists and half were art students. Suppose the average tourist took one half hour to view the collections while the average art student remained for two hours. A thoughtful museum guard made two observations. First, he noted that early in the morning the ratio of art students to normal patrons would be about 1 to 1. As the day progressed, the ratio of art students to tourists would gradually rise and then level off at 4 to 1. If this guard were to be brought into the museum not knowing the hour of the day, he could make a pretty good guess as to the time from the ratio of art students to tourists.

This same phenomenon goes on in our galaxy. Like the museum patrons, ^{235}U and ^{238}U enter at a regular rate. They remain for a time and then they leave via radiodecay. Because ^{238}U atoms hang around longer than do ^{235}U atoms, they are present in greater proportion in the galactic uranium pool than they are in the uranium that is added to the pool by supernovae.

The buildups of the ^{235}U, ^{238}U, and ^{232}Th contents of galactic matter are shown in Figure 4-5. Because of its short half-life, the amount of ^{235}U would level off several billion years after the galaxy formed. Hence, when the Earth formed, the ^{235}U content of the galaxy was very close to its steady-state value (that is, ^{235}U was decaying just about as fast as it was being produced). The amount of ^{238}U, with its 4.47-billion-year half-life, would still be climbing 10 billion years after the galaxy formed and would not level off for 10 billion or so years. ^{232}Th has a half-life about equal to the age of the universe. Thus, its amount in our galaxy is still only half the value that will be achieved many tens of billions of years from now. When the steady-state abundances are finally established in our galaxy, the abundances will be as follows (in this case, for clarity, the normalization has been made to 100 ^{235}U atoms instead of 1000 ^{232}Th atoms):

^{235}U	:	^{238}U	:	^{232}Th
100	:	410	:	2460

These steady-state values can be compared with the abundances in the solar nebula 4.6 billion years ago:

^{235}U	:	^{238}U	:	^{232}Th
100	:	345	:	818

Using this normalization, the ratios in fresh supernova debris are:

^{235}U	:	^{238}U	:	^{232}Th
100	:	66	:	127

Clearly, primordial solar matter did not have the proportions of these three nuclides that would be expected had element formation been going on for many tens of billions of years. Assuming that the ^{235}U amount had reached steady state 4.6×10^9 years ago, then ^{238}U had gotten 0.84 (345/410) of the way and ^{232}Th 0.33 (818/2460) of the way to steady state. Shown in Figure 4-6 are the ratios expected for these nuclides for various times after element production began (solid curve). These two curves cross dashed horizontal lines representing the values that existed at the time the Earth formed. The points at which the lines intersect should

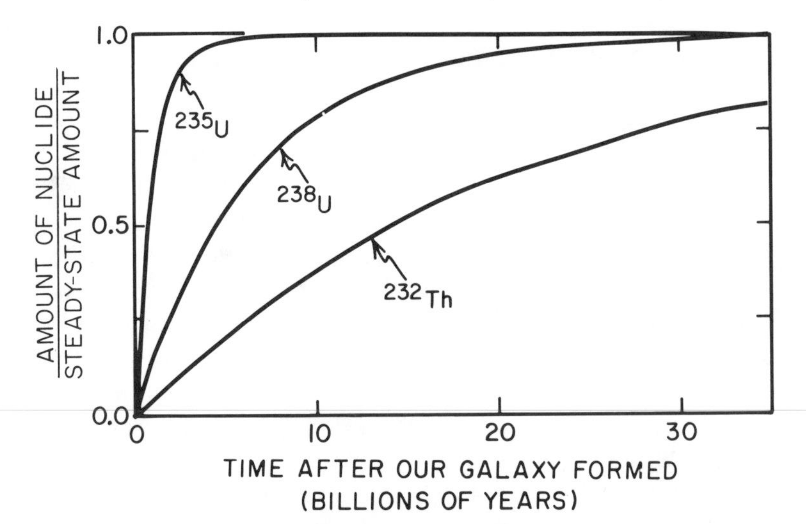

Figure 4-5. The evolution of the amounts of ^{235}U, ^{238}U, and ^{232}Th in our galaxy: The assumption is made that supernova events have occurred regularly over the entire history of our galaxy. The steady-state amounts correspond to the situation where the isotope is undergoing radiodecay at a rate that matches the new production in stars.

give the time elapsed between the formation of our galaxy and the formation of our solar system. For the $^{238}U/^{235}U$ plot, the crossover is at about 12 billion years. For the $^{232}Th/^{235}U$ ratio, the crossover is at about 9 billion years. While both results have rather large uncertainties, they do suggest that element production began about 10 billion years before our solar system came into being.

Combining the information from the red shift-versus-distance relationship with that obtained from radioisotopes, we get the scenario shown in Figure 4-7. The universe began about 15 billion years ago. Our galaxy formed sometime during the first billion years of universe history. Over our galaxy's entire history, red giants have steadily synthesized elements. The material in our solar system was isolated from the galaxy 4.6 billion years ago. While element synthesis has continued right up to the present, there is no mechanism by which the elements produced over the last 4.6 billion years could have become incorporated in significant amounts into the Sun or its planets.

Daughter Products of Extinct Radioactivities

Far more spectacular conclusions are forthcoming from observations involving the daughter products of now-extinct radionuclides. Two of these nuclides are particularly important. The first, ^{129}Xe, is the decay product of ^{129}I, which has a half-life of 16 million years (0.016 billion years). The other, ^{26}Mg, is the decay product of ^{26}Al, which has a half-life of 0.7 million years (0.0007 billion years). The half-lives of both of the parent nuclides are long enough that they persisted for an appreciable period of time after being spewed forth from a supernova. However, the half-lives are far too short to permit them to survive for the entire 4.6 billion years of solar-system history.

To understand the story these isotopes have to tell requires that we say a few things about isotopic abundances. In most cases the isotope abundance ratios for most elements is the same for all Earth rocks and meteorites. Where differences exist they are of two origins. The first has to do with radioactive decay. We have already seen that the isotopic composition of the element stron-

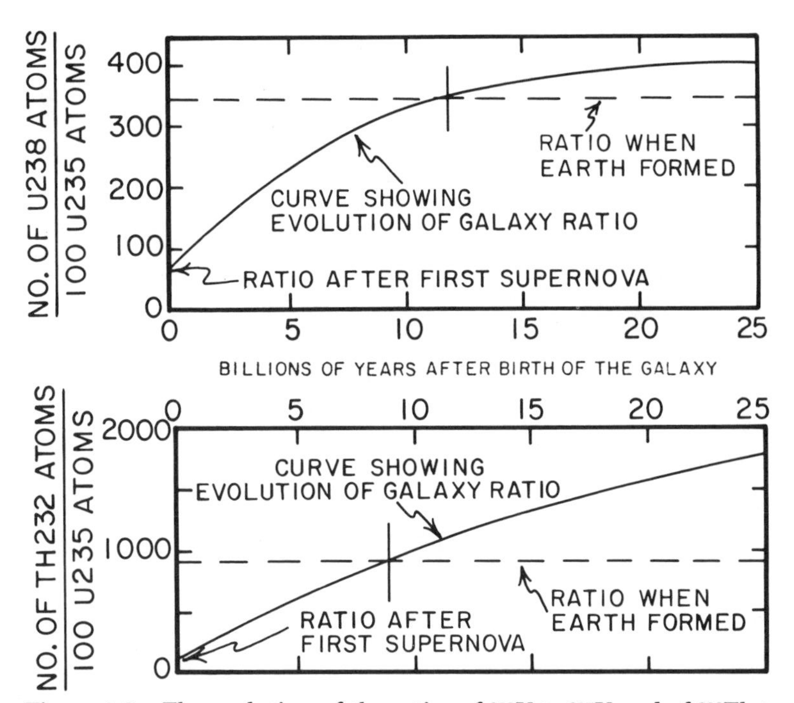

Figure 4-6. The evolution of the ratios of ^{238}U to ^{235}U and of ^{232}Th to ^{238}U in our galaxy if heavy-element production occurred at a constant rate: Very early in the galaxy's history the ratios were equal to the ratio in which they were produced in stars. With time the ratio changed favoring the longer-lived of the two isotopes. The horizontal dashed lines correspond to the ratios at the time the solar system formed. The intersection between the dashed line and the solid evolution curves should correspond to the time between the formation of our galaxy and the formation of our solar system. In the case of the ^{238}U-^{235}U pair, this time is about 12 billion years. In the case of the ^{232}Th-^{235}U pair, the time is about 9 billion years. Hence, about 10 billion years elapsed between galaxy and solar-system formation.

tium differs from mineral to mineral in the same meteorite. These variations have their origin in the radioactive decay of long-lived ^{87}Rb. The other source of isotope ratio differences has to do with small separations among isotopes during some chemical processes. $H_2{}^{16}O$ molecules evaporate just a bit more readily than do $H_2{}^{18}O$ molecules. $^{12}CO_2$ molecules are fixed by plants during photosynthesis just a bit more readily than are $^{13}CO_2$ molecules. These isotope separations owe their existence to the differences in mass between isotopes (that is, to the difference in the number

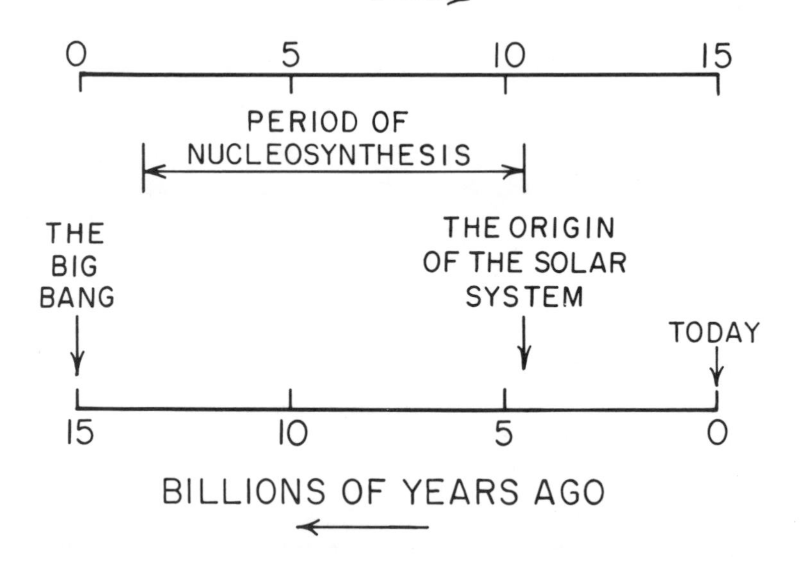

Figure 4-7. Summary of the chronology of universe events: The period of nucleosynthesis refers to the time interval over which the elements heavier than hydrogen and helium that are found in our solar system were produced. For the galaxy as a whole the period of nucleosynthesis extends right up to the present. The matter in the solar system was isolated from the galaxy 4.6 billion years ago.

of neutrons the respective isotopes contain). Despite the fact that the isotopes of an element have identical electron structures, the mass difference leads to small separations.* For elements with three or more isotopes these separations show a regular progression with mass (that is, isotopes differing by 2 in their neutron count are separated to twice the extent as are isotopes differing by 1 in their neutron number). Because of this mass dependence scientists are able to distinguish isotope anomalies produced

*The Manhattan Project carried out during World War II had as its most difficult and expensive task the separation of ^{235}U from ^{238}U. Even with the best brains in science and engineering and unlimited funds, the Project almost failed to produce enriched ^{235}U on schedule.

during chemical processes from those having their origin in radiodecay (at least, for elements having three or more isotopes).

Magnesium Isotope Abundance Anomalies

Jerry Wasserburg and his colleagues at the California Institute of Technology set out in 1974 to determine whether or not any radioactive ^{26}Al existed in the material from which meteorites formed. Their motivation came from a desire to learn about three aspects of the early development of the solar system:

1) Did a supernova event occur in our galaxy just before the solar system formed?
2) How long a time elapsed between the initiation of nebular collapse and the formation of meteorites?
3) Did heat from the decay of now-extinct radionuclides cause the objects present in the early solar system to melt?

Since ^{26}Al decays to ^{26}Mg the Wasserburg team launched a study to find whether anomalies in the relative abundances of magnesium separated from chondritic meteorites could be found (see Figure 4-8). They chose their minerals with great care. In particular, they sought mineral grains rich in the element aluminum and nearly free of the element magnesium. This requirement was particularly difficult to meet in that magnesium is one of the most abundant elements in meteorites. Fortunately, the Allende meteorite contains rare grains of a mineral called feldspar which has the formula

$$Ca\ Al_2\ Si_2\ O_8$$

The architecture of this mineral is such that there is no place where a magnesium atom can comfortably reside. Hence, these feldspar grains have a very high ratio of aluminum to magnesium.

By making very precise measurements of the ratio of ^{26}Mg to ^{24}Mg in the trace amount of magnesium present in these feldspar grains and comparing them with similar very careful isotope measurements made on magnesium contained in other more common mineral grains, Wasserburg and his co-workers were able to show that the $^{26}Mg/^{24}Mg$ ratio was slightly higher in the magnesium from the feldspar than in magnesium from bulk

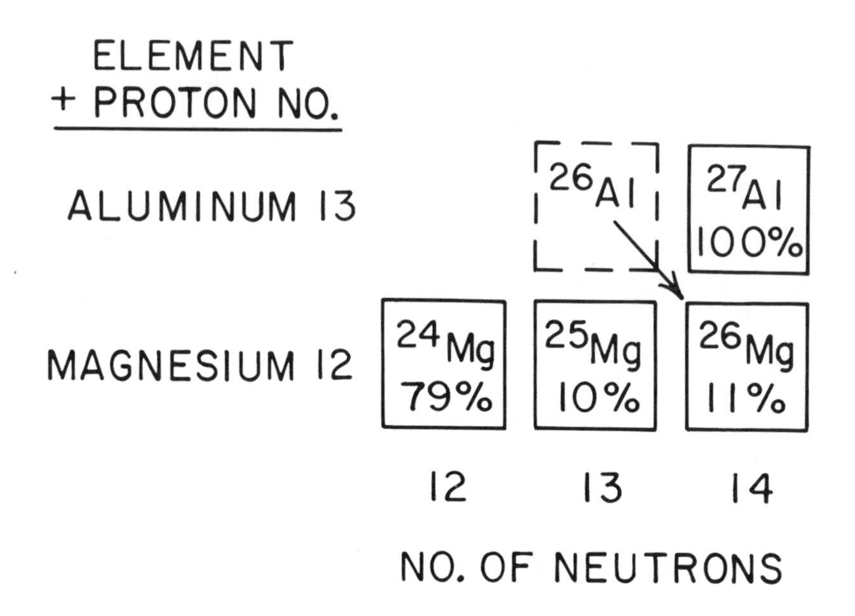

Figure 4-8. The isotopes of the elements aluminum and magnesium: Aluminum has only one stable isotope and magnesium three. Early in solar system history, however, a second aluminum isotope was present—radioactive ^{26}Al with a half-life of 0.73 million years. This isotope is no longer with us, having long since decayed to ^{26}Mg.

meteorites.

They were also able to demonstrate that this difference was not the result of chemical separations of the magnesium isotopes. They did this by carefully comparing the results for the ^{26}Mg/^{24}Mg ratio with those for the ^{25}Mg/^{24}Mg ratio. If the difference between the isotopic composition was the result of chemical separations, then the feldspar should have had a ^{25}Mg/^{24}Mg ratio anomaly equal to one-half ^{26}Mg/^{24}Mg ratio anomaly. It did not. Thus, Wasserburg et al. were confident that they were dealing with an isotope composition anomaly created by the decay of the now-extinct radioisotope ^{26}Al.

The Wasserburg team reasoned as follows. In a mineral that is rich in magnesium and poor in aluminum, the very tiny amount of ^{26}Mg produced by the decay of any ^{26}Al present when the mineral formed would be swamped out by the common ^{26}Mg (see Figure 4-9). However, in the feldspar, the ^{26}Mg produced by the decay of ^{26}Al would be subject to a far smaller dilution by common ^{26}Mg. Hence, the feldspar magnesium should have somewhat more ^{26}Mg atoms per gram of magnesium than average meteorite magnesium. This is what they found.

Another important crosscheck was made. Two different feldspar separates with different Al to Mg ratios were analyzed. As shown in Figure 4-10 a correlation exists between the aluminum to magnesium ratio and the ^{26}Mg to ^{24}Mg ratio. The more aluminum relative to magnesium, the larger the amount of ^{26}Mg in the magnesium. This is as expected if the excess ^{26}Mg

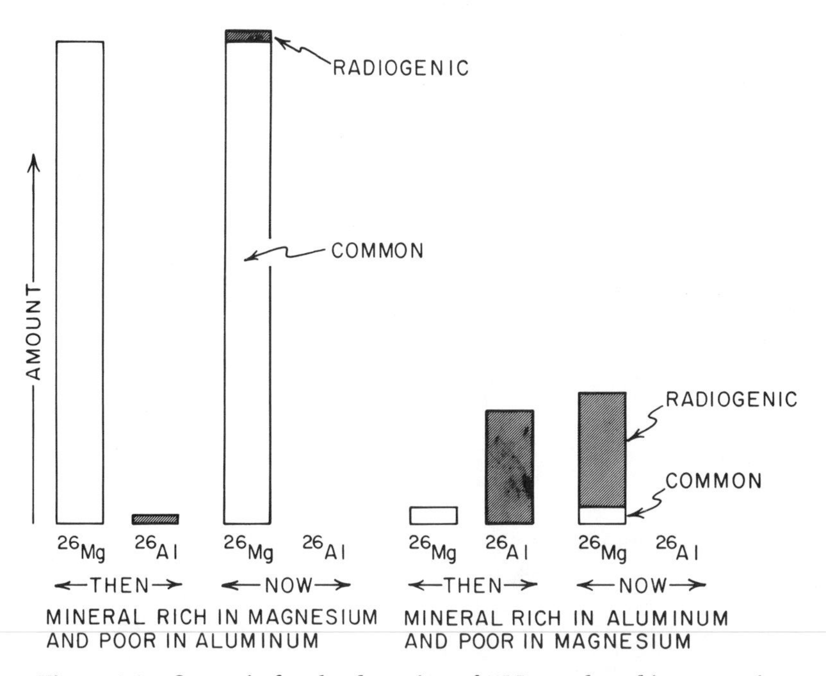

Figure 4-9. Scenario for the detection of ^{26}Mg produced in meteorites by the decay of ^{26}Al: To prove that ^{26}Al was initially present in meteorites, scientists had to search for the excess ^{26}Mg produced by ^{26}Al decay. As shown in these block diagrams the ratio of "radiogenic" ^{26}Mg to the "common" ^{26}Mg will be far larger in a mineral rich in the element aluminum and poor in the element magnesium than in one where the opposite situation exists. In order to make the point, the amount of ^{26}Al has been greatly magnified. In meteorites the amount of radiogenic ^{26}Mg is about a hundred times smaller than shown here!

was produced within the meteorite by the decay of ^{26}Al.

From this information these workers were able to calculate how many ^{26}Al atoms there were per unit of aluminum in the Allende meteorite when it formed. They obtained the following result:

$$\frac{^{26}Al}{Al} = 5 \times 10^{-5}$$

They compared this with the ratio calculated for fresh supernova debris:

$$\frac{^{26}Al}{Al} = 100 \times 10^{-5}$$

Here things got a bit dicey. ^{26}Al has a half-life of 0.73 million years. At one extreme we could envision that the time elapsed between the supernova explosion that produced the ^{26}Al and the time meteorites formed was so short that very little of the radioaluminum had a chance to decay. By "very short" we mean less than 0.1 million years (0.0001 billion years). In this case the last supernova event must have contributed 5 percent of the aluminum present in meteorites. This can be seen as follows. If 5 percent of the aluminum in the cloud that condensed to form the Sun was from fresh supernova debris with an $^{26}Al/Al$ ratio of 100×10^{-5} and 95 percent of the aluminum was from much earlier supernovae in which the $^{26}Al/Al$ ratio was zero, then the mixture of these two types of aluminum would have an $^{26}Al/Al$ ratio of 5×10^{-5}.

If, on the other hand, meteorites had formed about 1.4 million years after the supernova event, then only one-quarter the initial amount of ^{26}Al would remain. At this time the supernova aluminum would have had an $^{26}Al/^{27}Al$ ratio of 25×10^{-5} (as opposed to 100×10^{-5} for fresh supernova debris). In this case, in order to obtain the 5×10^{-5} value reconstructed for meteorites at the time of their formation, 20 percent of the aluminum would have to have come from the last supernova event and only 80 percent from the myriad of previous events.

Even a contribution of 5 percent of the heavy elements by this last event seemed to Wasserburg and his team to be on the high

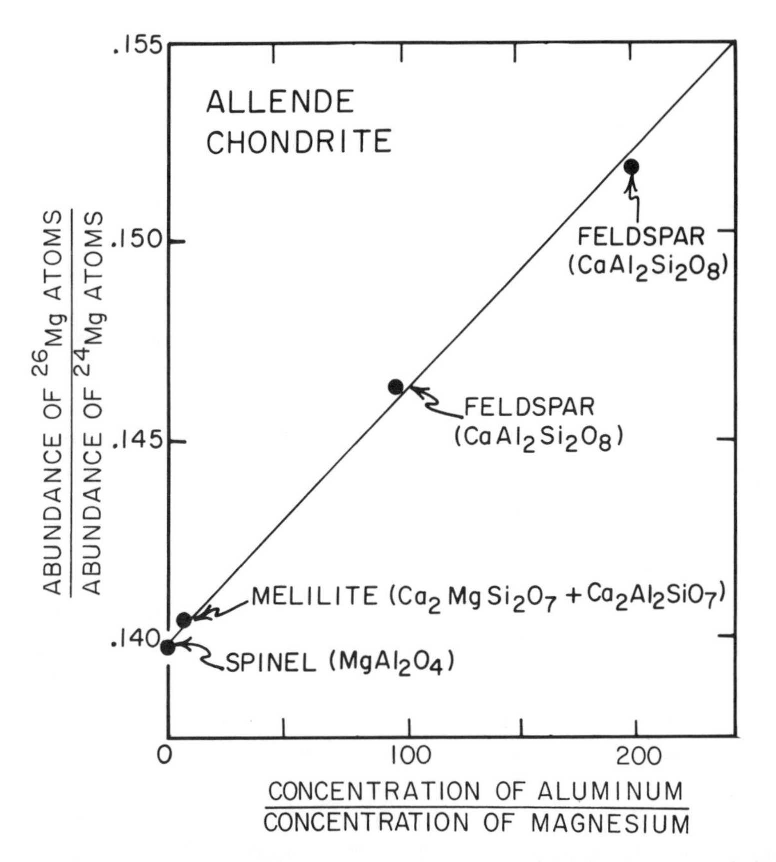

Figure 4-10. Relationship between the ratio of ^{26}Mg to ^{24}Mg and the ratio of aluminum to magnesium in mineral grains from a chondritic meteorite: As can be seen, feldspar grains which have aluminum as a primary constituent and magnesium as a trace constituent have more ^{26}Mg per unit of ^{24}Mg than do mineral grains where both aluminum and magnesium are primary constituents. The ratio of magnesium isotopes in average Earth matter is 0.1394. This is not surprising since the Al/Mg ratio in the Earth is about 0.1.

side. In a sense they had found too much of a good thing. They wanted to establish whether or not a supernova event had immediately preceded the solar system's formation. To do this they had to find evidence for the presence of short-lived ^{26}Al in newly formed meteorite mineral grains. Indeed, they found the desired evidence but it points to too much ^{26}Al. Something is not quite right.

Xenon Isotope Abundance Anomalies

Almost two decades before the Wasserburg team made their search for anomalies in the isotopic composition of the element magnesium, a team working at Berkeley under the direction of John Reynolds made a similar search for anomalies in the isotopic composition of the element xenon. Our reason for discussing the magnesium isotope results first is that they are more easily understood. The xenon isotope results lead to similar conclusions, but the reasoning involved is a bit more complex and the conclusions are not as crisp. It must be emphasized, however, that in discovering the isotopic anomalies in xenon, Reynolds and his associates established this powerful approach to the study of early events in our solar system.

One of the reasons why the situation for xenon is more complicated is that two extinct radioisotopes contributed to the differences found for its isotopic composition. When Reynolds launched his search he focused his attention on ^{129}Xe. His motivation was the possibility that the extinct radioisotope ^{129}I might have been present in meteorites when they formed. The Reynolds team found not only a very large difference between the ^{129}Xe content for xenon extracted from meteorites and the ^{129}Xe content for xenon from the Earth's atmosphere, but also smaller anomalies in the abundances of several other xenon isotopes. It was later shown that these additional anomalies were produced by the spontaneous fission of the now-extinct radionuclide ^{244}Pu (half-life 83 million years).

The first positive result obtained by the Reynolds group is shown in Figure 4-11. Since Reynolds did this pioneering work, xenon from a great many meteorites has been analyzed. Not only has the isotopic composition of the xenon been determined, but also the concentrations of both the elements xenon and iodine. As the excess ^{129}Xe must have been ^{129}I when the meteorite formed, the ^{129}I to ^{127}I ratio at that time can be computed. As shown in Figure 4-12, about one ^{129}I atom existed for each 10,000 ^{127}I atoms.

The isotope-abundance anomalies for xenon are generally considerably larger than those found for magnesium. As we have seen magnesium is one of the most abundant elements in meteorites. Thus, there is a lot of ordinary magnesium in which to hide that bit of ^{26}Mg produced by ^{26}Al decay. Xenon is a noble gas. Only a tiny amount of the available xenon was captured by meteorites. Although it is a moderately volatile element, iodine was captured with somewhat greater efficiency by the objects in the planetary system than was xenon. Thus the ^{129}Xe produced within the meteorites by the decay of ^{129}I had relatively little common xenon in which to hide and it produced comparatively large isotope composition anomalies.

Taken together, the magnesium and xenon isotope data seem to point to a heavy-element production event just prior to the birth of the solar system. They also require that the time between this event and the formation of meteorites be quite small.

When first discovered, this evidence was touted to support the idea that the shock wave from the supernova blast triggered the collapse of the quiescent cloud of dust and gas from which our solar system formed. However, the very large amount of ^{26}Al was a red flag. Astrophysicists found it hard to accept that a single supernova event would add more than a fraction of 1 percent of new material to its surroundings. The 5-percent figure seemed outrageous. The situation became even more complicated when the anomalies produced by other extinct radioactivities were measured—so complicated, in fact, that at the time this book was written no viable explanation for all the anomalies had been identified. Astrophysicists seem to have accepted the reality that there is more to the element-formation story than they had contemplated. Environments other than red giants and mechanisms other than those outlined above must be involved.

Isotope Anomalies for Oxygen

The story does not end here, however. Another type of isotope anomaly was found which can be attributed to neither the decay of any radioisotope nor to isotope separations during chemical reactions. Its discovery brought with it considerable confusion which has yet to be resolved to everyone's satisfaction. Bob Clayton, working at the University of Chicago, made a study of

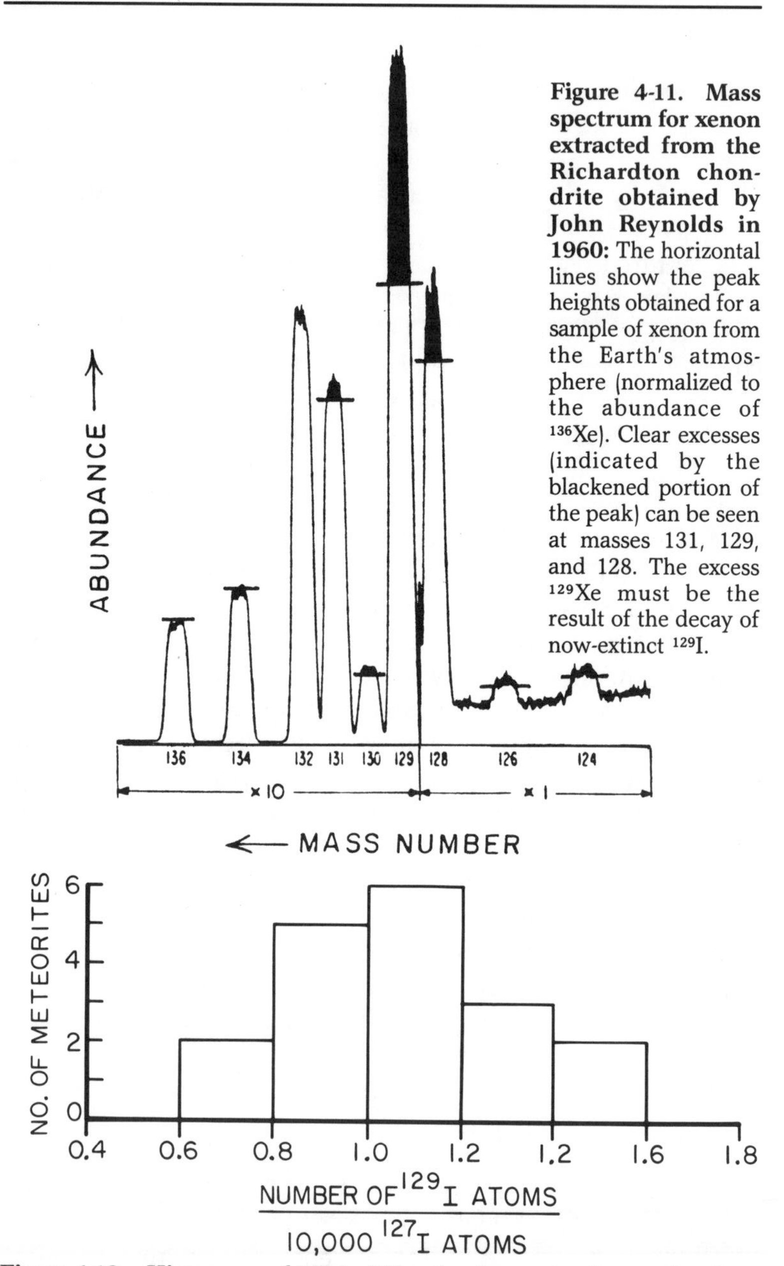

Figure 4-11. Mass spectrum for xenon extracted from the Richardton chondrite obtained by John Reynolds in 1960: The horizontal lines show the peak heights obtained for a sample of xenon from the Earth's atmosphere (normalized to the abundance of ^{136}Xe). Clear excesses (indicated by the blackened portion of the peak) can be seen at masses 131, 129, and 128. The excess ^{129}Xe must be the result of the decay of now-extinct ^{129}I.

Figure 4-12. Histogram of ^{129}I to ^{127}I ratios for meteorites at the time of their formation as reconstructed from measurements of excess ^{129}Xe: The excess ^{129}Xe found in the meteorite today was ^{129}I when the meteorites formed. Note that the total range of the ^{129}I to ^{127}I ratios is only a factor of two.

the abundances of the three stable isotopes of the element oxygen in mineral grains separated from carbonaceous chondrites. He found differences that were not mass–dependent. Clayton had previously shown that samples from a variety of Earth materials yielded ^{18}O abundance anomalies always just twice those for ^{17}O (as expected if oxygen isotopes were separated during chemical reactions). By contrast, in the mineral grains taken from carbonaceous chondrites he found that the percentage differences in ^{18}O and ^{17}O from ^{16}O were exactly the same. These patterns are shown in Figure 4-13.

Clayton was fully aware that he could not call on an extinct radioactivity to explain these results. The half-lives of any possible oxygen isotope precursors are measured in seconds or less. Hence, within minutes after a supernova no further oxygen was formed. Thus he realized he had discovered some other phenomenon. His explanation was that "alien" dust grains were present in the cloud which collapsed to form our solar system. He postulated that these alien grains originated from a star that blew out material with a mix of nuclides different than that for average galaxy matter. These grains drifted into one vicinity of the galaxy and were incorporated (without melting) into meteorites. In picking minerals for his analyses Clayton by chance encountered pieces of material that contained these alien grains.

While most scientists accepted both Clayton's results and also his explanation, it took a while for the full significance of this discovery to sink in. If grain to grain differences in isotopic compositions exist for oxygen then why not for magnesium and xenon as well? If so, then how many of the observed differences are attributable to the decay of now-extinct radioisotopes in newly formed meteorite grains and how many to the inclusion in these meteorites of "alien" dust grains with different isotope compositions?

Further, as melting or vaporization of the nebular material would have destroyed the identity of any alien grains, Clayton's results suggest that at least some of the material that collapsed to form the solar system survived melting throughout the entire period of solar-system formation. This finding, then, seem-

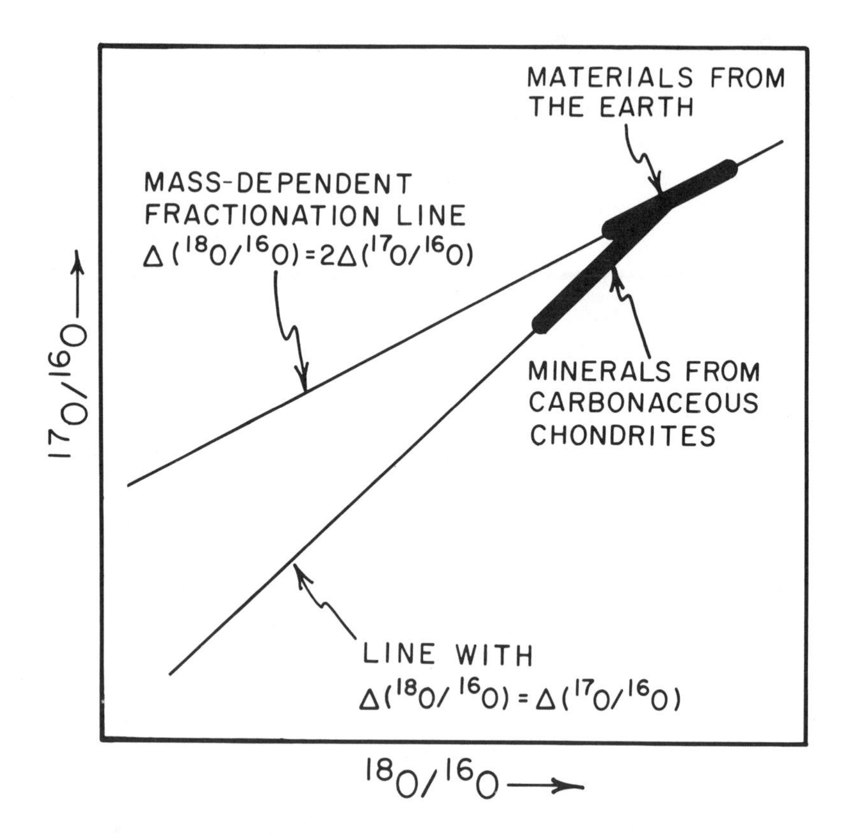

Figure 4-13. Diagrammatic representation of the abundance variations for the three isotopes of oxygen: While the variations on the Earth can all be explained by mass-dependent isotope separations occurring during chemical reactions, those in carbonaceous chondrite appear to be due to the inclusion of alien grains with an anomalous ^{16}O content (i.e., oxygen with the same $^{17}O/^{18}O$ ratio as Earth oxygen but more ^{16}O than Earth oxygen).

ingly eliminates from contention those hypotheses of planet origin that call, as many do, for condensation of solids from an ultra-hot cloud in which no solids were initially present.

Not liking to be forced into Clayton's box, scientists have squirmed to find a way out. One possible route has been found. Laboratory experiments carried out by Thiemands and Heidenreich at the University of California, San Diego, show that under certain very special circumstances, chemical reactions induced in gases by electron impacts can cause isotope separations of the type needed to explain Clayton's results. This finding added more confusion to an already muddled arena. We are talking about frontier science. Tantalizing facts have been gathered. It is not yet possible to say for sure what message these facts are trying to convey to us. Based on past performance, however, it is likely that the scientists will sort things out. Perhaps, by the turn of the century, the true message carried by the rapidly growing zoo of isotope anomalies will be deciphered!

Summary

In this chapter we have attempted to show how the decay of certain long-lived radionuclides can be harnessed to tell time. The first example is a success story. Through studies of abundances of parent and daughter isotope pairs it has been possible to show that the mineral grains in chondritic meteorites crystallized 4.6 billion years ago. These results also suggest that the formation of the solar system was completed in a period not exceeding a few million years. Thus, the age of 4.6×10^9 years applies to the Sun and applies to the planets as well as to meteorites.

The second example did not lead to such a crisp answer. In this case, the ratios of long-lived isotopes were used to estimate the time of formation of the heavy elements. The analysis shows that the elements that make up the Earth and meteorites on the average predate our solar system by several billion years. To be more specific requires a knowledge of how the rate of formation of red-giant stars has changed over the course of galactic history. Hence, we can only show that the present-day abundances of the long-lived isotopes of uranium and thorium are consistent with a more-or-less constant rate of red-giant formation over a period starting about 10 billion years before our solar system formed.

While no firm conclusions can yet be drawn from the third example,the prospects are very exciting. Studies of the abundances of isotopes produced by the decay of now-extinct radioisotopes ^{26}Al, ^{129}I, and ^{244}Pu strongly suggest that an injection of new elements was made to our neighborhood of the galaxy just before the solar system came into being. If this proves correct then two important insights emerge. First, the time required to produce a star and its planets once the collapse is initiated is by universe standards very small. To explain the ^{26}Mg data these events must occur in less than the half-life of ^{26}Al (0.73 million years). Second, the radioactivity of the ^{26}Al incorporated into early planetary material provided a potent source of heat which could have caused the large objects in the solar system to melt very early in their history.

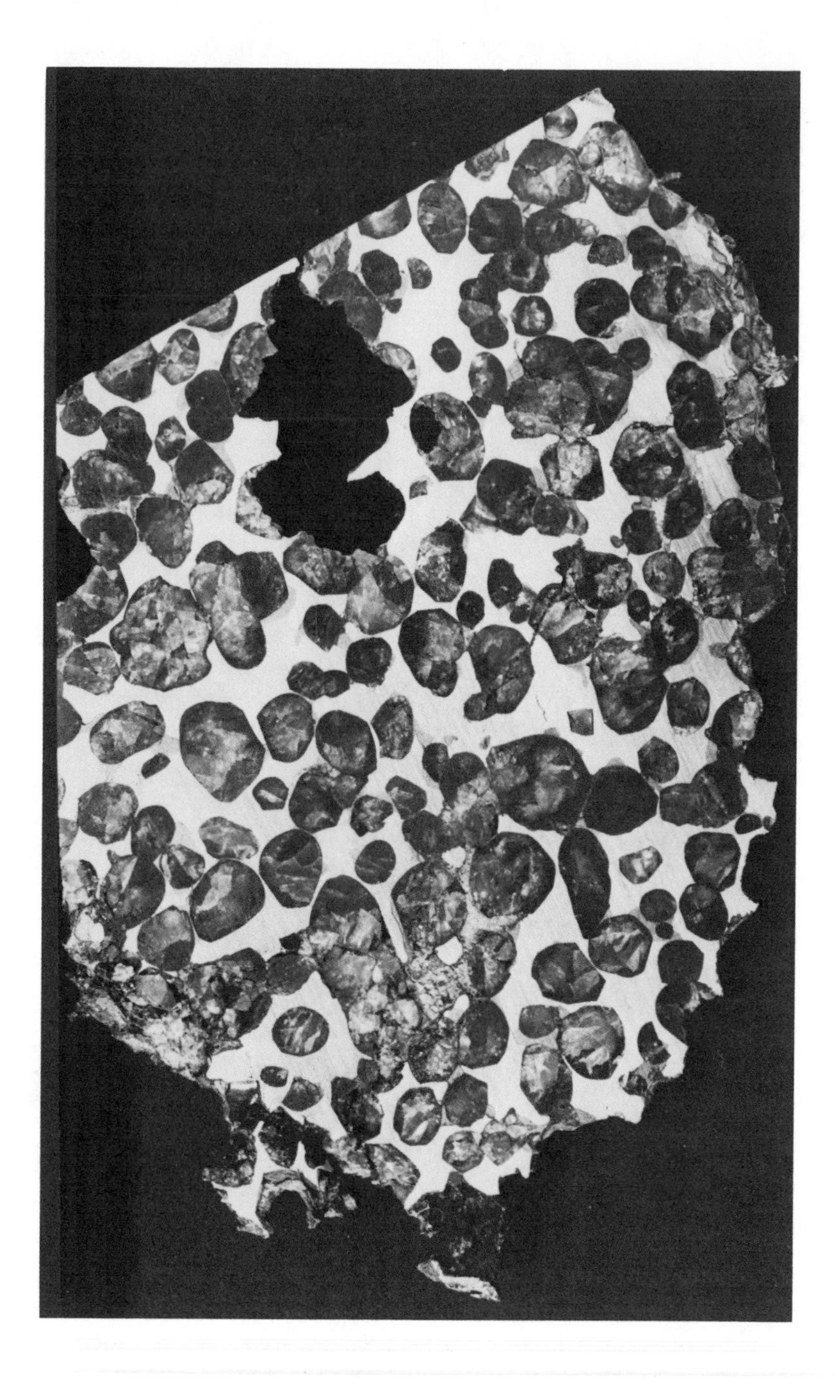

Stony Iron Meteorite

Chapter Five

Interior Modifications:

Segregation into Core, Mantle, Crust, Ocean, and Atmosphere

The main constituents of the meteorites that fall on the Earth's surface are oxides of silicon, iron, and magnesium and metallic iron. These objects are pieces broken loose during collisions between objects in the asteroid belt. They provide a model for the structure and chemical composition of the Earth and its fellow terrestrial planets. In chondrite-bearing meteorites the metal and silicate phases are randomly mixed on a millimeter scale. This is taken as the initial form for the material ultimately built into the planets. By contrast, in those meteorites where chondrules are absent, the silicate and metal phases have become segregated, as shown here by the polished section of the Brenham meteorite. From this we get the idea that the asteroids from which these meteorites originated melted, allowing the dense metallic iron to trickle to the core and the silicate to wrap itself around this core as a mantle.

Although no one has visited our Earth's interior, we have ample evidence from the manner in which earthquake waves pass through the body of our planet, from its bulk density, and from its magnetic field that Earth too has an iron core and a silicate mantle. Isotope clocks suggest that the separation into core and mantle was accomplished very early in Earth history. It is likely that the volatile element compounds that make up the Earth's atmosphere and ocean were purged from the Earth's interior at the time of core formation.

By contrast, the Earth's crust has formed and reformed over the entire history of the planet. The thin basaltic crust, which lies beneath the oceans, is replaced entirely once each 0.1 billion years. The thick continental crust has accumulated over the last 3.8 billion years of Earth history. Crustal formation is a by-product of the slow convective flow of the solid mantle which carries to the Earth's surface heat released by the radioactive decay of uranium, thorium, and potassium atoms contained in the mantle.

Introduction

Unlike chondrule-bearing meteorites which have survived the 4.6 billion years of solar-system history without melting, most material in the planetary system appears to have been melted one or more times. In the case of the terrestrial planets, these episodes permitted a segregation into layers differing in chemical composition. This process has had a phenomenal impact on the environment of the surface of the Earth.

The main impetus for the segregation of material in planets comes from the great difference between the density of iron metal on one hand and that of silicate on the other. Just as rocks fall to the bottom of the sea or oil floats to the top of water, the iron metal in a planet should migrate to its center. While the tendency for this segregation exists, it will not happen as long as the object remains cold. If the object melts, however, segregation will promptly occur. Thus, as our first step in considering the chemical reorganization of material in planets, let us see what evidence we have for the pooling of iron at their centers.

We learned in Chapter Three about one prominent type of meteorite: objects consisting entirely of iron-nickel metal alloys. The existence of these objects strongly suggests that at least some of the asteroids melted. If a carbonaceous chondrite were to be melted in the laboratory, two immiscible liquids would form. One would be dominated by metallic iron. The other would be dominated by oxides of magnesium, iron, and silicon. The metal liquid, being more dense than the oxide liquid, would trickle to the bottom of the container. Similarly, if an asteroid were to have melted, liquid metal would have settled to its center to form a core, and the liquid oxide would have wrapped itself around this core to form a "mantle." When it once again solidified, the asteroid would be chemically layered. Iron meteorites are presumably chips from the cores of such asteroids, and basaltic achondrites (that is, meteorites with no chondrules and no metallic iron) are presumably chips of their mantles.

Evidence for an Earth Core

Taking advantage of the pulses of sound made by earthquakes, seismologists have been able to determine the internal structure of our planet (see Figure 5-1). The most important finding made in this way is that a profound discontinuity exists at a depth of 2900 kilometers. Using seismic data in concert with other information, geophysicists have been able to show that this discontinuity divides the materials making up the Earth into two distinct categories.

The material above the discontinuity transmits both the compressional and shear waves generated by earthquakes, while the material below it transmits only the compressional waves. This tells the seismologist that the material above the discontinuity is solid while that below is liquid. Further, the velocity at which compressional sound passes through the material above the discontinuity is considerably greater than the velocity at which this sound passes through the material below the discontinuity. This tells the seismologist that the material below the discontinuity is more malleable than that above the discontinuity. Through detailed analyses of data from many earthquakes received at stations all around the Earth it has been possible to construct a plot of density versus depth in the Earth (see Figure 5-2). By comparing these densities with those observed during laboratory squeezing tests, the geophysicists have been able to show that the densities for the material above the discontinuity are consistent with those expected for magnesium silicate (minerals consisting of the oxides of magnesium and silicon), and that those for the material below the discontinuity are consistent with the densities expected for metallic iron at the pressures existing in the Earth's core. Thus, the obvious conclusion is drawn: like many of the asteroids, the Earth must have a magnesium silicate mantle and a predominantly iron core. Unlike the metal core in an asteroid, that in our planet is still molten.

Many checks on this hypothesis are available. Pieces of mantle rock thrust up through the crust to where we can get at them have been shown to have chemical compositions similar to those for basaltic achondrites. The Earth's magnetic field can be

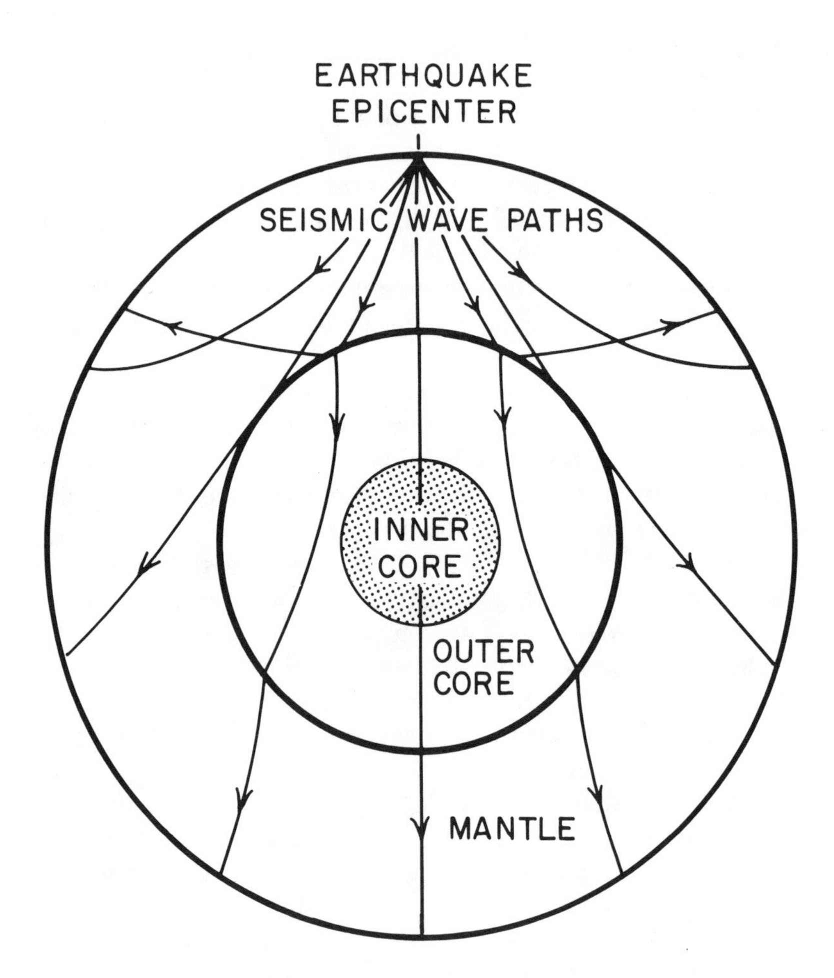

Figure 5-1. Seismic waves provide data on the physical properties of the core: Two types of waves move through the Earth from the epicenter of an earthquake: compressional (waves that move back and forth in the direction of their travel) and shear (waves that move at right angles to their direction of travel). Shear waves do not travel through liquids, while compressional waves do. The fact that only compressional waves pass through the core demonstrates that at least the outer core is liquid. Besides information regarding the physical state of material in the core, seismology yields information regarding the distribution of density with depth in the Earth, greatly aiding in the establishment of the bulk chemical composition of the Earth's mantle and core.

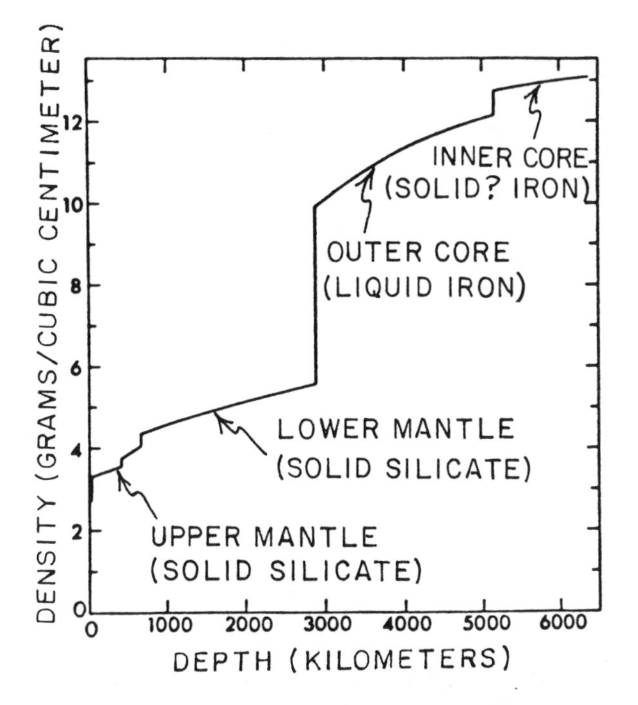

Figure 5-2. Density as a function of depth in the Earth: The density at Earth's center (6300 km) is about 13 grams/cubic centimeter. Just beneath the core-mantle boundary (2900 km) it is 10 gm/cm³. The discontinuous change in density from 10 gm/cm³ to 5.5 gm/cm³ at 2900 kilometers is thought to represent the boundary between a liquid iron core and a solid silicate mantle. The gradation of density within the core and within the mantle is due to the increase with depth of the weight of the overlying material. In the absence of gravitational squeezing liquid iron has a density of 7.5 gm/cm³ and solid silicate a density of about 3.3 gm/cm³.

explained if eddies of molten metal are present in its deep interior (see Figure 5-3). At the pressures that exist at 2900-kilometers depth, metallic iron would melt at a considerably lower temperature than magnesium silicate. Thus, there is no problem in having liquid iron abutting solid silicate. The Earth's shape is oblate, that is, the poles are slightly flattened and there is a slight equatorial bulge. While this basic shape is that expected for any spinning planet, the magnitude of the flattening can only be explained if the Earth's mass is concentrated toward its center.

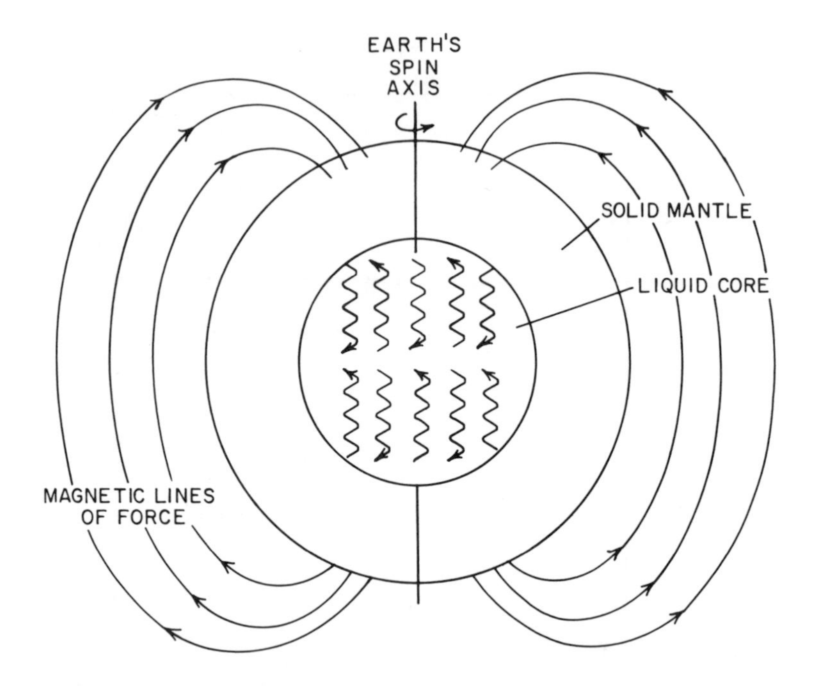

Figure 5-3. The Earth's magnetic field is generated by a liquid iron dynamo in the core: While the details of how the dynamo works are not known, in the model depicted here it is assumed that the electrically conducting metallic liquid of the core flows in screw-like rollers.

So convincing is this combined evidence that there is universal acceptance of the basic Earth structure described here.

Before moving on to the other planets, it is worthwhile to pause and see how the iron content of the Earth given in Table 3-6 of Chapter Three was obtained. The mass of the Earth's core is estimated by geophysicists to be 1.87×10^{27} grams. That of the mantle is estimated to be 4.11×10^{27} grams. Together the masses of the Earth's thin outer crust, ocean, and atmosphere come to only 0.01×10^{27} grams. The iron content of mantle material is estimated from measurements on lavas coming from deep in the Earth to be about 8 percent by weight. The iron content of the core is estimated to be 85 percent by weight (the rest presumably is nickel with perhaps some sulfur). From these results the iron content of the planet can be estimated:

mass of iron in core	1.6×10^{27} gm
mass of iron in mantle	0.2×10^{27} gm
total mass of iron	1.8×10^{27} gm
total mass of Earth	6.0×10^{27} gm

This yields an iron content for average Earth material of 30 percent.

Do Venus and Mercury Have Cores?

While Venus and Mercury have corrected densities high enough to demand that they contain sizable amounts of metallic iron, no success has been had in proving that this iron has accumulated at the centers of these planets. One way to find out would be to use space probes to deploy seismographs on the surfaces of these planets. "Planetquakes" would eventually supply evidence which would allow an analysis similar to that done for Earth. No attempt has as yet been made to do this. Indirect information could conceivably be obtained by looking for magnetic fields surrounding these planets and by checking the extent to which their poles are flattened. No magnetic field has been observed for either Venus or Mercury, nor has the polar flattening of these planets been determined with sufficient accuracy. But because neither Venus nor Mercury spins at an appreciable rate (as listed in Table 1, Chapter Three, they have very long "days"), these tests are not definitive. Both the generation of eddies in a planet's core and the significant flattening of its poles require a considerably faster spin than that of Venus or Mercury.

The Chemical Side Effects of Iron Cores

A number of other elements readily dissolve in liquid iron metal. Nickel has chemical affinities for metallic form even stronger than those of iron. Thus, although nickel is as abundant as calcium and aluminum in the Earth as a whole, most of the Earth's nickel is out of our reach in the core. Rocks found at the Earth's surface are low in nickel. Only a few key nickel-bearing ore deposits have been found to supply industry's demand for this metal. Gold, silver, and platinum also were depleted from the mantle by core formation, making these already rare elements even more precious. As we shall see in Chapter Six, because the Earth's iridium resides largely in its core, it has been possible to show that a large asteroid or comet hit the Earth 60 million years ago. Finally, much of the Earth's sulfur is probably safely stored in the Earth's core as iron sulfide. This is a fortunate circumstance, for many of the common chemical compounds of sulfur are toxic to man.

The fact that 85 percent of the Earth's iron is stored in the core as iron metal also makes a major difference to the mineralogy of the Earth's mantle and crust. Our planet would be quite different if its iron were instead in oxide form. To understand this difference, we must consider the minerals that exist in the mantle. When mixtures of SiO_2, MgO, and FeO are melted in the laboratory, depending on the proportions of these three oxides in the mix, three different mineral assemblages can form. One is a mixture of the minerals olivine and pyroxene. Both of these minerals accommodate the elements iron, magnesium, silicon, and oxygen. Olivine has the chemical formula

$$(Fe, Mg)_2SiO_4$$

and pyroxene has the chemical formula

$$(Fe, Mg)SiO_3$$

Each mineral has sites available for ions with two plus charges (Mg^{++} and Fe^{++}). Iron or magnesium atoms substitute for one another in these sites. Each mineral has sites available for ions with four plus charges (Si^{++++}). Only silicon can occupy this site. Both minerals have sites available for ions with two minus charges ($O^{=}$). Only oxygen can fill these sites. As all matter must be electrically neutral in each of these minerals, the number of negative charges must balance the number of positive charges. If a melt is to crystallize to form olivine and pyroxene crystals, then its ratio of Fe + Mg atoms to Si atoms must lie between 1 (that found in pyroxene) and 2 (that found in olivine). In this case, these two minerals together could use up all the available ingredients.

A second possible mineral assemblage is pyroxene plus quartz. The chemical formula for quartz is SiO_2. The four positive charges on the silicon atom are balanced by the minus charges on the two oxygen atoms. If the melt is to crystallize to form this mineral pair, then the ratio of Fe + Mg to Si atoms would have to be less than unity. In this situation, when all the Fe and Mg atoms had been built into pyroxene there would be silicon atoms left over. They would form quartz.

A third possible mineral assemblage is olivine plus wustite. The chemical formula of wustite is FeO. In this mineral there is a site for an iron atom with two plus charges and a site for an oxygen atom with two minus charges. If the melt is to crystallize to form this mineral pair, then the ratio of Fe + Mg to Si atoms would have to be greater than 2. In this situation, when all the silica and magnesium had been built into olivine, iron atoms would be left over. They would form wustite.

For the Earth as a whole, the ratio of Fe + Mg to Si lies just above 2. Thus, if *all* the iron were in oxide form, then the major mineral would be olivine. Wustite would be present as well. The olivine would contain nearly equal numbers of iron and magnesium atoms. On the other hand, if all of the iron were in the form of metal, then the Fe + Mg to Si ratio in the oxide material would be just above unity and the dominant mineral would be pyroxene. This pyroxene would be rich in magnesium and poor in iron. In addition, there would be a small amount of olivine.

Examples of this mineralogic difference are seen in the varieties of chondritic meteorites. In a variety called enstatite chondrites the iron is almost entirely in the metal and sulfide forms. No olivine is found in these meteorites. The dominant silicate mineral is pyroxene. The pyroxene has a high Mg to Fe ratio. By contrast, in another variety of chondritic meteorites, the so-called olivine pigeonites, only about 20 percent of the iron is in metal and sulfide form. In these meteorites the dominant silicate is olivine. The olivine has about 35 percent iron and 65 percent magnesium atoms.

The game in Figure 5-4 helps us to understand this concept.

Given some mixture of the four elements iron, magnesium, silicate, and oxygen, the player is asked to find which three of the five possible mineral phases are present. He also is asked to find what proportion of each element is in each of these three mineral phases. Thus we see that the presence of metallic iron in a planet not only permits it to develop a core but also alters the mineralogy and trace-metal chemistry of its mantle. These differences leave their imprint on the composition of the planet's outer crust.

Mechanisms for Core Formation

Like most of the processes involved in planet formation, those leading to core formation are subject to controversy. At one pole in this controversy stand those who propose that the Earth grew in layers; the metallic iron accumulated first and then the oxides. At the other pole are those with the more conventional view that the Earth formed as a mixture of metal and oxide material; subsequently the object melted and became segregated.

Let us consider first the conventional view. If the Earth was formed of a homogeneous mixture of metallic iron and silicate material, then at some time after it formed it must have melted. Only if it were molten could the iron have segregated readily to the planet's center. In this case, the main question to be answered is what caused the Earth (and other objects in the planetary system) to melt. There are two sources of energy which could have done the job: the gravitational energy released as the material making up the Earth accumulated, and the energy released by the decay of the radioactive isotopes present in the accumulating material. The first of these sources is certainly adequate. Were all the gravitational energy released during planet accumulation to have been retained by the material in the planet, then the material in Mercury would have heated by 4100° C, that in Mars by 5900° C, that in Venus by 25,000° C, and that in Earth by 29,000° C.* At the surface pressures of Earth the metallic iron melts at 1535° C. While the melting temper-

*The more massive the planet, the greater the energy released during an object's impact.

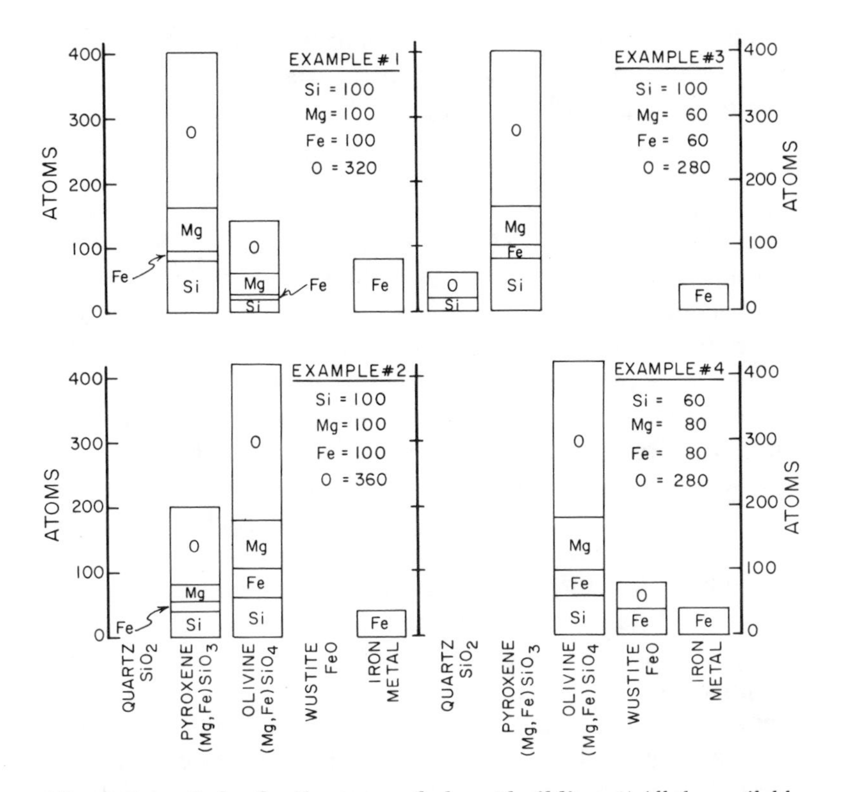

Figure 5-4. Rules for the game of planet building: 1) All the available atoms must be used to construct only three of the five possible mineral phases; 2) while the proportions of atoms in a given mineral must be exactly those stated in its formula, the ratio of iron to magnesium in the olivine and pyroxene can be set at any value.

atures of iron are a thousand or so degrees higher at the pressures existing in the Earth's interior, the energy supplied by impacts is still more than adequate to do the job. The problem lies in the assumption that all this heat was retained. Since the impacts of individual planetesimals onto a growing planet would, of course, have been at its surface, the heat generated could have been radiated away to space before the impact craters were buried beneath the debris of subsequent impacts. To assess how much of the impact energy was lost to space and how much was held within the growing planet requires a knowledge—

which we don't have—of the details of the sizes of the impacting planetesimals and the rates of their accumulation.

As discussed in Chapter Four, we have evidence that now-extinct radioactive nuclides were present when chondrites formed. If these isotopes were also built into the planets, then the heat released through their radiodecay could have released enough heat to drive the planet's temperature up by several thousands of degrees. In this case the timing of planet formation becomes very important. Since ^{26}Al has a half-life of only 0.73 million years, its potency as a heat source dropped rapidly with time. If more than several millions of years elapsed between the formation of meteorites and the time the planets accumulated, then ^{26}Al's punch would have been lost.

Now let us look at the other scenario. If iron condensed first from the nebula to produce metallic iron protoplanets around which layers of magnesium silicate later were added, then there should be a clear difference between the temperature at which iron metal condenses into solid form and that at which magnesium silicate condenses into solid form. As shown in Table 5-1, no such clear difference exists. If the nebular density lay near the best-guess value, then the condensation of iron metal and the condensation of the magnesium silicate should have proceeded together. On the other hand, if the density of the cloud from which the Earth condensed was a factor of 10 higher than the best guess made by planetary scientists, iron would have condensed at a slightly higher temperature than silicate (that is, in a cooling nebula, iron would have condensed first).

Because of our lack of firm knowledge regarding the conditions that prevailed at the time the planets formed, we cannot distinguish between these two hypotheses. The question remains open.

The Timing of Core Formation

If our planet did accumulate as a heterogeneous mixture of metallic iron and magnesium silicate, and if either heating by the gravitational energy associated with impacts or heating by nuclear energy associated with the radioactivity of ^{26}Al suc-

Table 5-1. Temperature of condensation of solids in a nebula with a hydrogen gas to a pressure of 10^{-4} atm (i.e., the best-guess value).

Corundum	Al_2O_3	≃1680° K
Spinel	$MgAl_2O_4$	≃1440° K
Iron metal	Fe	≃1360° K
Olivine	Mg_2SiO_4	≃1350° K
Pyroxene	$MgSiO_3$	≃1300° K
Feldspar	(K,Na) $AlSi_3O_8$	≃1000° K
Iron sulfide	FeS	≃ 650° K

ceeded in making the planet hot enough to allow the iron metal to segregate, then core formation must have occurred right at the beginning of Earth history. However, if instead we call on the heat produced by the decay of the long-lived radionuclides ^{40}K, ^{238}U, ^{235}U, and ^{232}Th to do the job, then at least several tenths of a billion years would be required to raise Earth's temperature to the melting point. In this case the core would have formed well along in the history of the Earth rather than at the very beginning.

Here again the daughter products of the long-lived radioisotopes have something to tell us. That something is that the Earth's core very likely formed quite soon after the Earth formed. One source of information is the trend in the isotopic composition of the element lead through geologic time resulting from the decay of ^{238}U, ^{235}U, and ^{232}Th. As this argument involves grasping several difficult concepts, we will only state its gist. The idea is that, like nickel, much of the lead present in the Earth must have gone to the core along with the iron. By contrast, uranium remained behind in the mantle. Thus, the core-formation event produced a sudden large decrease in the ratio of lead to uranium in the Earth's mantle. In so doing it strongly influenced the evolution through time of the mantle abundance of ^{206}Pb and ^{207}Pb (produced by the decay of uranium) relative to that of ^{204}Pb (not produced by the decay of any long-lived nuclide). Measurements of these ratios in meteorites and in lead minerals from very ancient Earth rocks suggest that the separation of lead and uranium must have occurred during the first 0.1 billion years of Earth history.

The other argument is more easily grasped, so we will say a

bit more about it. In 1983 Claude Allegre and his co-workers in Paris produced convincing evidence that the ^{129}Xe content of xenon associated with recent volcanic eruptions is significantly higher than the ^{129}Xe content of xenon from our atmosphere (see Table 5-2). Since the source of volcanic lavas studied by the Allegre team was the Earth's mantle, this difference indicates that the decay of the now-extinct radionuclide ^{129}I contributed less ^{129}Xe to the xenon now in the atmosphere than to the xenon now in the Earth's mantle. Since ^{129}I has a half-life of only 16 million years, these differences must have their origins very early in Earth history, that is, in the first 0.1 billion years.

The Allegre team sees these results as evidence for early core formation. The argument they put forth runs as follows. If the Earth formed after the last of the ^{129}I in the nebula had decayed, then all of its xenon should be isotopically uniform. The fact that differences are found in the ^{129}Xe content of xenon gas from different sources requires that some ^{129}I was built into the Earth.

Allegre and his colleagues envision that the Earth not only formed but also melted before its ^{129}I had decayed away. This melting allowed two things to happen. It allowed the iron metal to migrate to the Earth's center, and it allowed much of the gas

Table 5-2. Isotopic composition of xenon gas contained in recent basalts: The abundance of each xenon isotope is referenced to 1000 atoms of ^{130}Xe. The basalt xenon has a clear excess of ^{129}Xe over that for atmospheric xenon while no difference is seen in the relative abundances of the other four isotopes.

Source of Xenon	^{128}Xe	^{129}Xe	^{130}Xe	^{131}Xe	^{132}Xe
Earth's Atmosphere	470	6480	1000	5190	6590
Hawaiian basalt	455±10	6600±20	1000	5250±20	6590±30
Atlantic Ocean basalt	459±15	7230±40	1000	5200±30	6660±90
Indian Ocean basalt	469±20	6710±20	1000	4820±20	6620±20
Pacific Ocean basalt	470±10	6810±70	1000	5210±60	6620±70
Galapagos basalt	486±21	6790±70	1000	5180±50	6650±60
Basalt / Atmosphere	No difference	Basalt higher	—	No difference	No difference

trapped in the body of the planet to migrate to the surface. It is difficult to see how one of these separations could occur without the other happening as well. The Earth's iodine would have stayed with the mantle material. Hence, this event would have separated a nearly homogeneous planet into core, mantle, and atmosphere. Most of the iron in the Earth would have gone to the core, and most of the xenon in the Earth would have gone to the atmosphere. Most of the iodine (including any ^{129}I still around) would have stayed in the mantle.

The Earth then cooled to the point where a solid crust formed, isolating the mantle from the atmosphere. Once this had happened, any ^{129}Xe produced by the decay of the remaining ^{129}I would have been trapped in the mantle, further enhancing the ^{129}Xe content of any xenon that remained behind.

A scenario patterned after that envisioned by Allegre is shown diagrammatically in Figure 5-5. Between the time of the Earth's formation and the time of the core formation event (taken quite arbitrarily in this scenario to be a period of 32 million years), three-quarters of the ^{129}I initially present in the Earth would have decayed. The ^{129}Xe produced from this decay would have been distributed uniformly in all the Earth's xenon. The scenario further assumes that during the core-formation event 80 percent of the Earth's xenon escaped to the surface and became part of the atmosphere (like that of 32 million years, the choice of 80 percent is arbitrary). The other 20 percent of the xenon failed to escape and remained in the mantle. The ^{129}Xe produced by the ^{129}I still present after this event was added to the residual xenon in the mantle over the next few tens of millions of years. This addition changed the isotopic composition of mantle xenon. During this time the Earth's atmosphere was isolated from the mantle by a solid crust which prevented the mantle's residual xenon from escaping to the atmosphere. Thus, the isotopic composition of the xenon in the atmosphere did not change significantly. In this way two reservoirs of xenon were generated, each with its own ^{129}Xe content.

In fairness to the supporters of built-in Earth layering, it is quite possible that our mantle was produced mainly by the in-

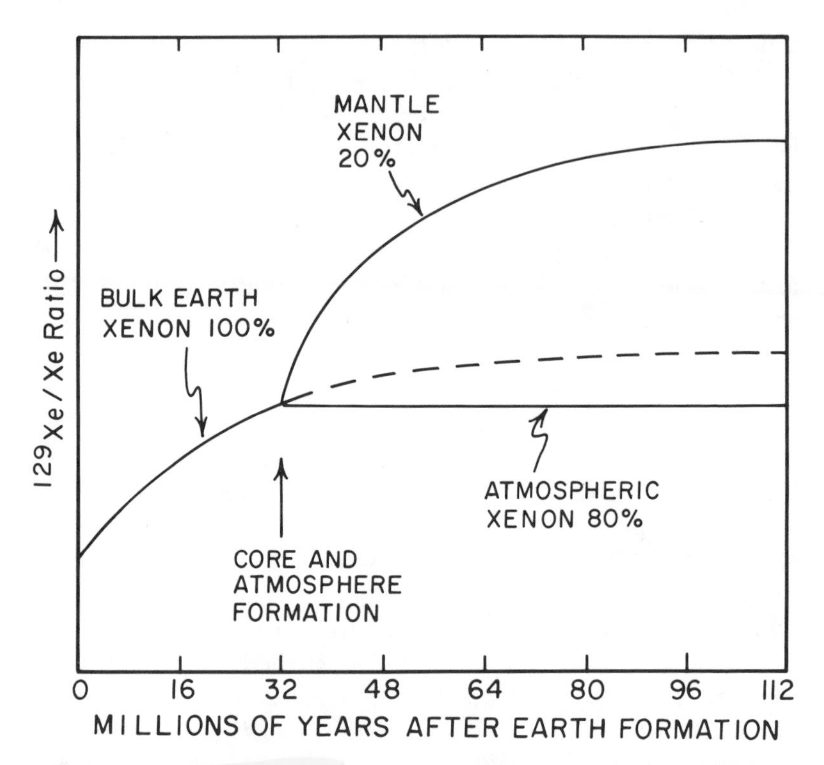

Figure 5-5. Hypothetical scenario for the evolution of the ^{129}Xe content in Earth xenon: At the time the Earth formed it was a homogeneous body made of small mineral grains such as those seen in chondrites. Its xenon was trapped in the mineral grains. With the passage of time the ^{129}Xe content of the Earth's xenon was increased through the decay of ^{129}I. Then at some point the Earth melted sufficiently to allow core formation. This event also permitted much of the gas initially trapped in mantle minerals to migrate to the Earth's surface. From that point in time (taken here quite arbitrarily as 32 million years after the Earth formed), there were two xenon reservoirs, the atmosphere and the mantle. As iodine remained in the mantle, the decay of the remaining ^{129}I enriched the ^{129}Xe content of only that xenon remaining in the mantle. The dashed line shows the evolution which would have occurred had the Earth not undergone core and atmosphere formation until well after all ^{129}I had decayed.

fall of nebular material from the Earth's vicinity, while the atmosphere was derived from the infall of late-arriving material originating in the region of the major planets. Perhaps the xenon in these two types of material had different ^{129}Xe contents. If so, the xenon isotope difference seen by the Allegre team may have been built in rather than generated by the core-formation event.

The Earth's Outer Crust

The point has already been made that the chemical composition of the rocks upon which we walk is quite different than that of the mantle. Like an apple, the Earth has a thin skin which is unlike the stuff that is underneath. In probing these differences and their origins, we take our first step toward understanding why the Earth provides a habitat so ideally suited for the development of life. It is here that our planet turns the corner away from the monotonous path dictated by element abundances and element volatility!

To appreciate this, let us consider the situation that would confront colonists arriving from afar to look over a solar system that had gone no further in its chemical evolution than had the asteroids. They would land on a rocky surface pockmarked by impacts of the late-arriving planetesimals. The rocks would consist mainly of magnesium, silicon, and oxygen (the planet's iron is in metallic form at its center). There would be no air, no water, no oil, no ore deposits, no life. If the colonists chose to stay they would face boredom and hardship. Nearly everything except crude shelters of hewn rock would have to be either imported or derived with great effort from the bulk rock.

Earth is different. It must be admitted that much of Earth's beauty and perhaps also its equitable environment are directly attributable to life itself. However, even if life had never taken hold, a colonist from space would still find Earth a much more desirable home than the planet described above. From here on we will emphasize those aspects of the Earth which gave rise to the wondrous setting it provides for life. We will see that the important factors are size, distance from the sun, rate of spin, and the degree to which the more volatile elements present in the solar nebula were captured.

Our first step will be to consider the origin of the part of the Earth's crust we call the continents. Although we perceive continents to be that portion of the Earth's surface emerging above sea level, the distinction is far more profound. If the ocean's waters were removed so that we could take vacations driving on highways built through its basins, we would see a far different landscape than we are accustomed to seeing on the continents. Geologic studies reveal that in addition to standing lower than the continents, the ocean-basin portion of the Earth's outer skin is fundamentally different in chemical composition from the continental portion of its skin. While the continents are constructed largely of rocks called granites, the ocean floor is constructed largely of rocks called basalt. As shown in Table 5-3, granite has a chemical composition far different from that for the Earth's mantle. The elements aluminum and sodium, which play minor roles in the mantle, assume primary roles in granitic crust. Even the element potassium, which is just a trace constituent of the mantle, becomes a significant mineral-maker in continental crust. Iron and magnesium are far lower in abundance in granites than in the mantle. By contrast, the basaltic crust of the ocean basins is chemically more akin to the mantle.

A look at the ages of these rocks reveals another startling difference.* The ocean's basalts rarely exceed 0.1 billion years in age,

Table 5-3. Chemical composition (in percent by weight) of the two most important rock types in the Earth's crust compared to the composition of Earth's mantle and to the composition of chondritic meteorites.

	Chondritic meteorites	Earth's mantle	Basalt	Granite
O	32.3	43.5	44.5	46.9
Fe	28.8	6.5	9.6	2.9
Si	16.3	21.1	23.6	32.2
Mg	12.3	22.5	2.5	0.7
Al	1.4	1.9	7.9	7.7
Ca	1.3	2.2	7.2	1.9
Na	0.6	0.5	1.9	2.9
K	0.1	0.02	0.1	3.2
Other	5.9	1.7	2.7	1.6

*The age of an igneous rock is the time since it crystallized from a liquid; for a metamorphic rock it is the time elapsed since its baking; for a sedimentary rock it is the time since its deposition. All these times are based on radioisotope hourglasses.

that is, most were formed in the most recent 2 percent of the Earth's history. The granites of the continents, on the other hand, range back in age to about 3.8 billion years.

Thus, when we talk of the Earth's outer skin we must distinguish the young, low-standing basalts of the sea floor from the old, high-standing granites of the continents. Clearly they are two different kettles of fish.

Volcanoes

Before considering the origins of these two distinct kinds of crust, let us consider the implication of the observation that molten rock still pours forth from the Earth's interior. The eruptions of active volcanoes (see Figure 5-6) provided man's first clue to the fact that the Earth's mantle is still very hot. To melt the materials found in these lavas requires temperatures in excess of 1000° C! Why, after 4.6 billion years, hasn't the Earth's interior cooled down?

The answer is that the Earth's interior is still being heated by the energy released through the radiodecay of four long-lived radioisotopes, ^{40}K, ^{235}U, ^{238}U, and ^{232}Th. This heat maintains the solid mantle of our planet at a high temperature and drives giant convection cells in order to escape. Geophysicists insist that the material making up the mantle is solid. Like Silly Putty, which shows an elastic response when bounced on the sidewalk, mantle material rings to the jolt of an earthquake. But, also like Silly Putty which seeps into the rug when left unattended, mantle material flows when subjected to small long-term stresses.

These motions of the material in the mantle form huge "convection cells." Material rises toward the crust, flows laterally beneath the crust, and then sinks, as shown in Figure 5-7. The convection cells carry this heat to the upper portions of the mantle. From here it works its way to the surface by conduction (as heat gets through a brick wall) and with waters recirculating through cracks in the crust. Once at the surface the heat is radiated into space. The amount of heat arriving at the surface is about 40 calories per square centimeter per year. While it is 6000 times less than the amount of heat reaching us from the

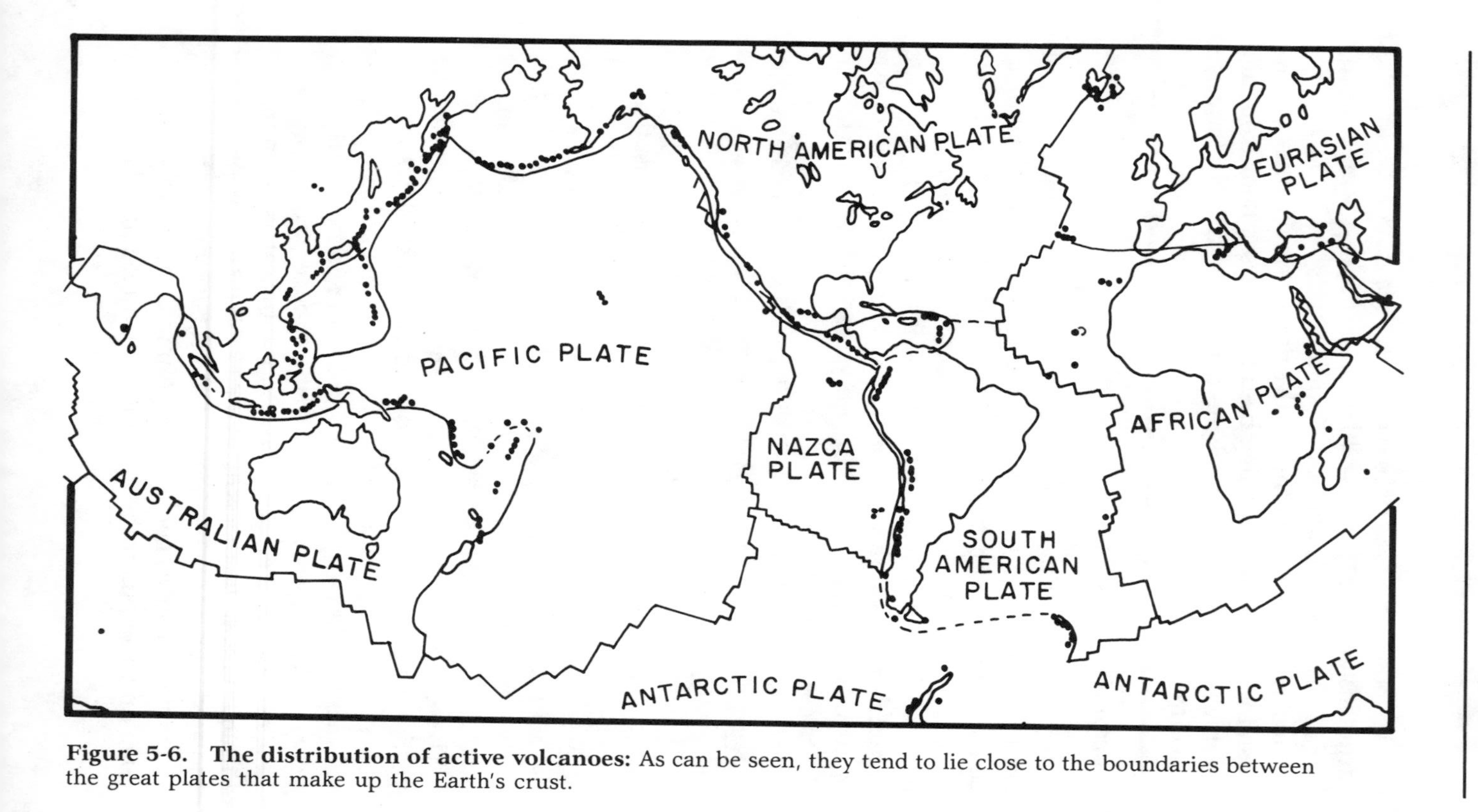

Figure 5-6. The distribution of active volcanoes: As can be seen, they tend to lie close to the boundaries between the great plates that make up the Earth's crust.

Sun, this small amount of radioactive heat is responsible for keeping the Earth geologically alive.

In order for the radioactive heat to be able to work its way through the material lying above the mantle's convection cells, the material in the uppermost mantle must be close to its melting point. So close in fact, that here and there blobs of molten material are produced beneath the crust. Some of these blobs break through the crust and spill out upon the surface. Others solidify within the crust before reaching the surface. Regardless of their fate, these mantle-generated liquids are ultimately responsible for creating and maintaining the Earth's outer skin. Active volcanoes remind us that this construction and repair work still goes on. By contrast, activity of this type long ago ceased in the asteroids and in Mercury. These objects are geologically dead.

Ocean Basins

The crust we associate with ocean basins has its origin along the zones where mantle material moves toward the surface (see Figure 5-7). As the hot material from deep in the mantle moves upward, the pressure exerted by the weight of overlying material drops correspondingly. This pressure drop reduces the temperature required for mantle material to melt.* Thus as the solid material pushes its way up, a little "juice" appears in the interstices of the rock. By the time the material has reached the base of the crust this liquid has increased in amount, reaching perhaps 15 percent of the total. Being less dense than the companion solid material, this liquid separates and rises up into the overlying crust. Here it is funneled through a great long crack created in the ocean crust by the divergent flow of neighboring mantle cells (see Figure 5-7). It fills the ever widening void, crystallizing to form new oceanic crust. This process is much the same as when ice forms on a cold winter day in a gap created when two ice blocks drift apart in a river. New ocean crust forms

*The melting points of almost all materials found on and in the Earth increase with increasing pressure. The main exception is ice.

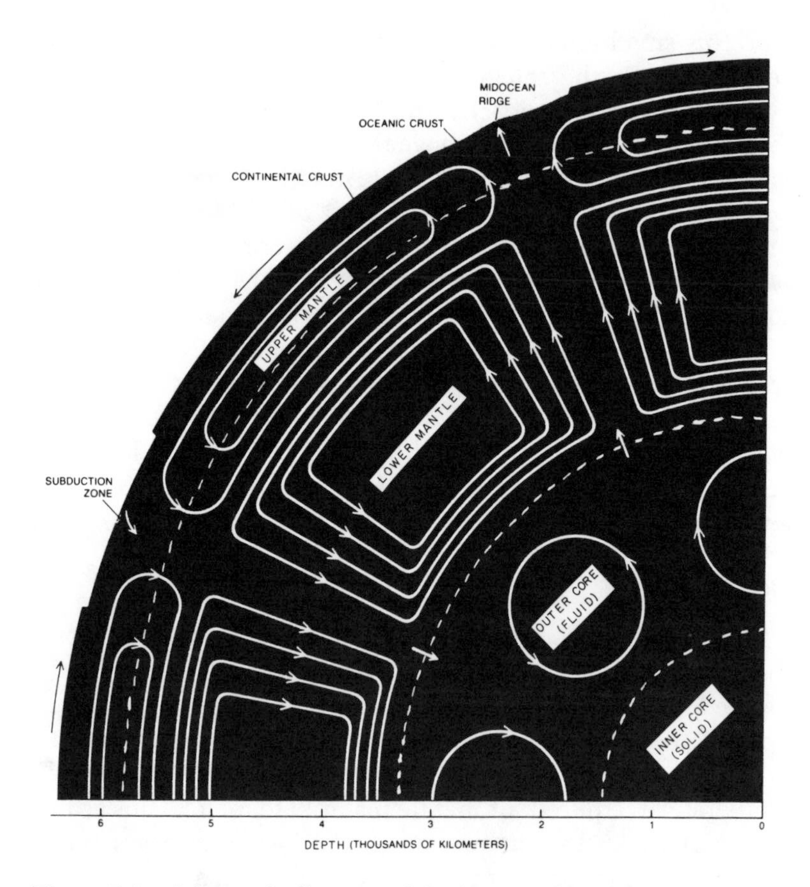

Figure 5-7. Schematic diagram of the large-scale motions within the Earth: Heat-driven convection in the fluid outer core has a dynamo effect that is responsible for the geomagnetic field. Convection in the upper mantle drives plate tectonics. Volcanism transports molten material to the surface at mid-ocean ridges and other places. Solid material is returned to the interior at subduction zones.

along these cracks at the rate of several centimeters per year.

There is a marked compositional difference between the liquid that rises to the surface and the residual solid material that remains behind in the mantle. Certain of the minerals in mantle material melt at lower temperatures than do others. Hence the chemical composition of the material that becomes liquid is not the same as that of the material that remains solid. Furthermore, each element has the opportunity to partition itself between the liquid and the solid. This partitioning is in accord with the chemical characteristics of the particular element. While the detailed rules governing this partitioning are sufficiently complex that not even the experts understand them all, the most important aspects can be described quite simply.

In Chapter Three the point was made that the element-to-element differences in the size of atoms are small. It was this near-constancy that allowed us to deduce something about the chemical composition of a planet from its bulk density. Now we shall show that although the range in size is small, it is great enough to play an important role in the partitioning of ions between melt and residual solid.

In Figure 5-8 are given the sizes of the ions of a number of the elements. They are arranged in accord with the chemist's periodic table (see Figure 5-9). Two trends are apparent. First, ions with the same charge (columns in the table) show an increase in size with increasing element number (down the columns). Second, ions with a single electronic structure (rows in the table) show a decrease in size with increasing positive charge (to the right). The upshot of this is that the ions of K^+, Na^+ and Ca^{++} are larger than the ions Mg^{++} and Fe^{++} which are in turn larger than the ions Al^{+++} and Si^{++++}.

Linus Pauling, the Nobel Prize-winning chemist, many years ago noted some simple regularities in the architecture of oxide minerals. Pauling's rules state that the negatively charged oxygen ions in minerals are arranged in polyhedra around the positively charged metal ions (see Figure 5-10). The number of oxygens in such a polyhedron must be as large as possible with the restriction that the metal ion at the polyhedron's center be large enough to keep the oxygens from touching one another (that is, each oxygen must lean on the central metal ion rather than on another oxygen ion). Using some simple trigonometry, Pauling calculated that a Si^{++++} ion could hold apart four oxygen ions (in a tetrahedron) but not 6 oxygen ions (in an octahedron). Mg^{++} and Fe^{++} ions could hold apart 6 oxygen ions (in a octahedron) but not 8 oxygen ions (in a cube). He then

ALKALI METALS	ALKALINE EARTHS					HALOGENS	NOBLE GASES
3 Li+ 0.78	4 Be++ 0.34				8 O= 1.32	9 F− 1.33	10 Ne —
11 Na+ 0.98	12 Mg++ 0.78	13 Al+++ 0.57	14 Si++++ 0.39	15 P+++++ 0.35	16 S= 1.74	17 Cl− 1.81	18 Ar —
19 K+ 1.33	20 Ca++ 1.06		26 Fe++ 0.82			35 Br− 1.96	36 Kr —
37 Rb+ 1.49	38 Sr++ 1.27					53 I− 2.20	54 Xe —

Figure 5-8. Ionic radii for the elements of main interest to us: Except for iron they are arranged as in the chemist's periodic table. The element number is given in the upper left corner of each box. The pluses and minuses indicate the electrical charge of the ionic form of the element common for the Earth's crust and mantle. The radii are given in units of 10^{-8} cm.

pointed out that ions too rare in abundance to form minerals of their own would tend to substitute for an ion in its own size class. This is also the reason that Mg^{++} and Fe^{++} ions freely sit in for one another in the minerals olivine and pyroxene.

Pauling's rules help us to understand which elements tend to be concentrated in the melt and which in the solid phase. Large ions like K^{+} and Na^{+} find no comfortable slots in the minerals of the mantle. Hence, when given the choice they opt for residence in the liquid. Thus, partial melting of mantle rock leads to an enrichment in the Earth's crust of large-sized elements! This concept was first emphasized by the late Paul Gast. Since it was Paul who first taught me the excitement of Earth science, I cannot let the opportunity slip by without mentioning his key role in the development of thinking concerning planetary chemistry!

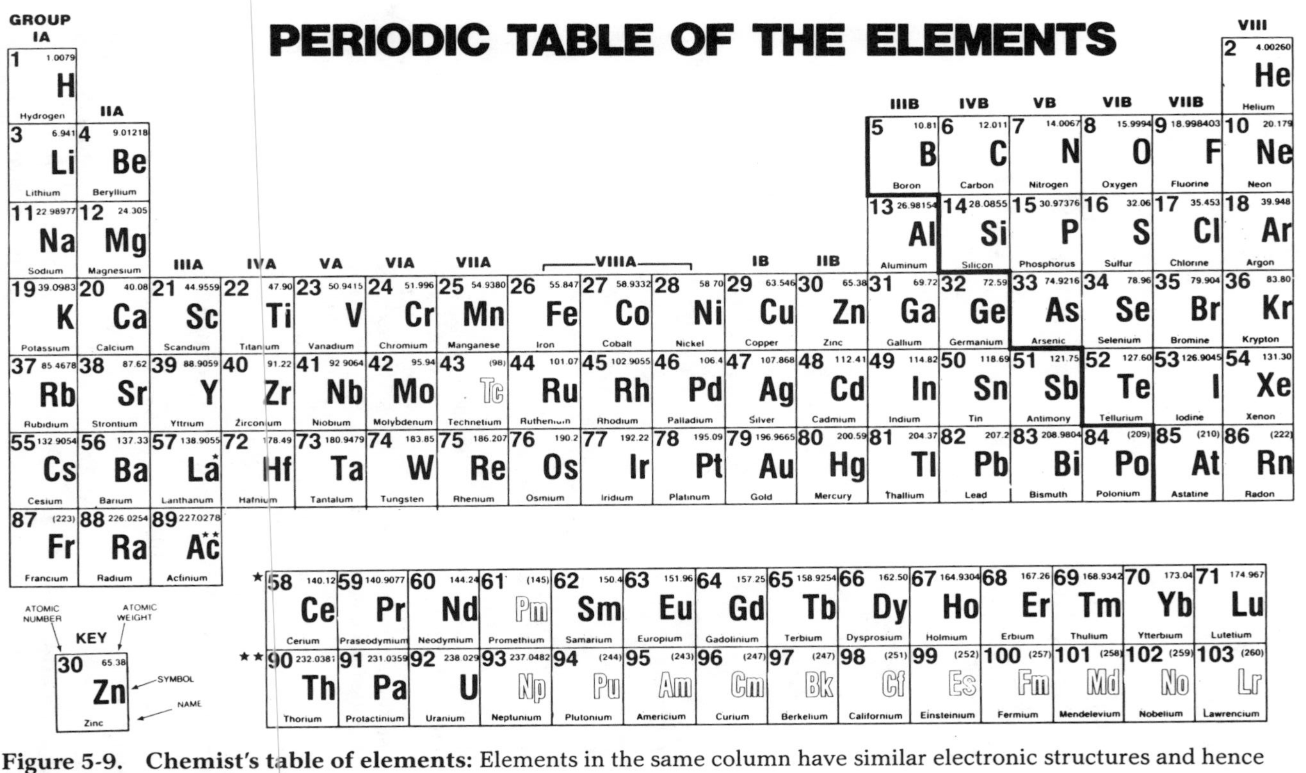

Figure 5-9. Chemist's table of elements: Elements in the same column have similar electronic structures and hence similar chemical tendencies.

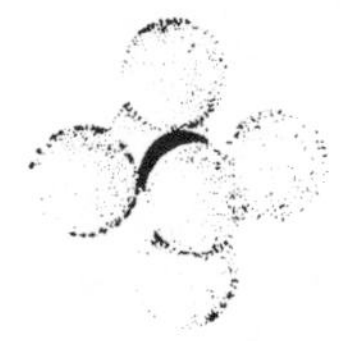

Figure 5-10. Coordination polyhedra made by oxygen ions around the metal ions in silicate minerals: The smallest cations lie between three oxygens; the carbon atoms in the mineral calcite are so situated. The next category of cations are those in tetrahedral coordination; the silicon atoms in all silicate minerals are so situated. The next category of cations are those in octahedral coordination; this is the preferred situation for magnesium and iron atoms in silicate minerals. Finally, large ions like potassium, sodium, and calcium prefer to be surrounded by even larger numbers of oxygens. One example of such an arrangement is the cubic coordination shown above.

Production of Andesite and Granite

The process of chemical refinement does not end with the formation of oceanic crust. A second stage occurs at the other end of the limb of the convection cell that carries the crust away from the oceanic spreading centers. As shown in Figure 5-7, at the place where the limbs of two neighboring cells converge, crustal material is driven down into the mantle. Most of the world's volcanoes overlie these zones of convergence (see Figure 5-6). Most geologists suspect that melting occurs preferentially along these zones because water is carried down with the crustal material.

As fresh oceanic crust cools during its journey away from the spreading center, minerals become "hydrated"* by interactions with the waters that move through cracks. As the "hydrated" crust later moves down into the mantle it gets heated by contact with the surroundings. The "hydration" reactions are turned around, releasing the mineral-bound water. In the presence of water, the melting temperatures of most silicate rocks plummet. Hence, even though the environment for these downward-penetrating convection-cell limbs is somewhat cooler than that for the rising limbs, melting can still occur. The liquids so formed rise and spout forth from volcanoes. The rocks so formed are called andesites. The composition of these rocks is one step further along the enrichment sequence from mantle material to granite.

The subsequent steps to granite are not well understood. We do know, however, that constituents of granite are the first liquids to appear when many types of rock are heated in the laboratory. In this sense granites are the ultimate product of repeated partial melting.

Elevation Differences

Another of the prominent contrasts between continental crust and ocean crust remains to be explained: the elevation difference. Both crusts float in the mantle much as wood floats on water. For the same reason that convection cells can operate in the solid mantle, the mantle adjusts to the differing weights exerted by neighboring units of crust. The situation is akin to that which would exist if we covered part of a swimming pool with

---

*By hydration is meant the incorporation of water molecules into the mineral structure. An example of a hydration reaction is

$$MgO + H_2O \rightarrow Mg(OH)_2$$

4 × 4 hardwood beams and part with 8 × 8 softwood beams. The softwood beams would float higher for two reasons: they are thicker and they are less dense. Similarly, the granite crust, which is thicker and less dense than the basaltic crust, floats higher in the mantle (see Figure 5-11).

The large difference in the thickness of the two crustal types owes its origin to the processes which create the crusts. The thickness of the ocean crust depends on the amount of liquid produced on the ascending limbs of the convection cells. It turns out that enough liquid is made to fill the opening void created by the separating plates to a depth of about 5 kilometers. Hence, the ocean crust has this thickness.

For the same reasons we had to be vague about how granites form, we must be vague about the factors controlling the thickness of the continental crust. The continents might be viewed as a scum which has built up on the churning mantle. It is buffeted about by the mantle's convection cells and pushed into ever thicker mounds. The thicker the mounds, the higher they float. The higher they float, however, the faster they erode. It is likely that a dynamic balance has been struck, such that the snowplow action of colliding plates, which piles the "scum" ever higher, is being countered by erosion which thins these piles.

It should be mentioned in this regard that if a continental block were to become any thinner than about 23 kilometers it would be covered by the sea. At this point erosion would cease. Because of this the continents are everywhere at least 23 kilometers in thickness. The thickest continents lie beneath the highest mountains. Here the crust achieves thicknesses approaching 60 kilometers. These portions of the continent float so high that they are capped with permanent glaciers. The movement of these glaciers grinds down these rising peaks at a fearsome rate. Thus, the continental crust thickness is limited on one side by the cessation of erosion when it drops beneath the sea and on the other by the acceleration in erosion rate caused by mountain glaciers.

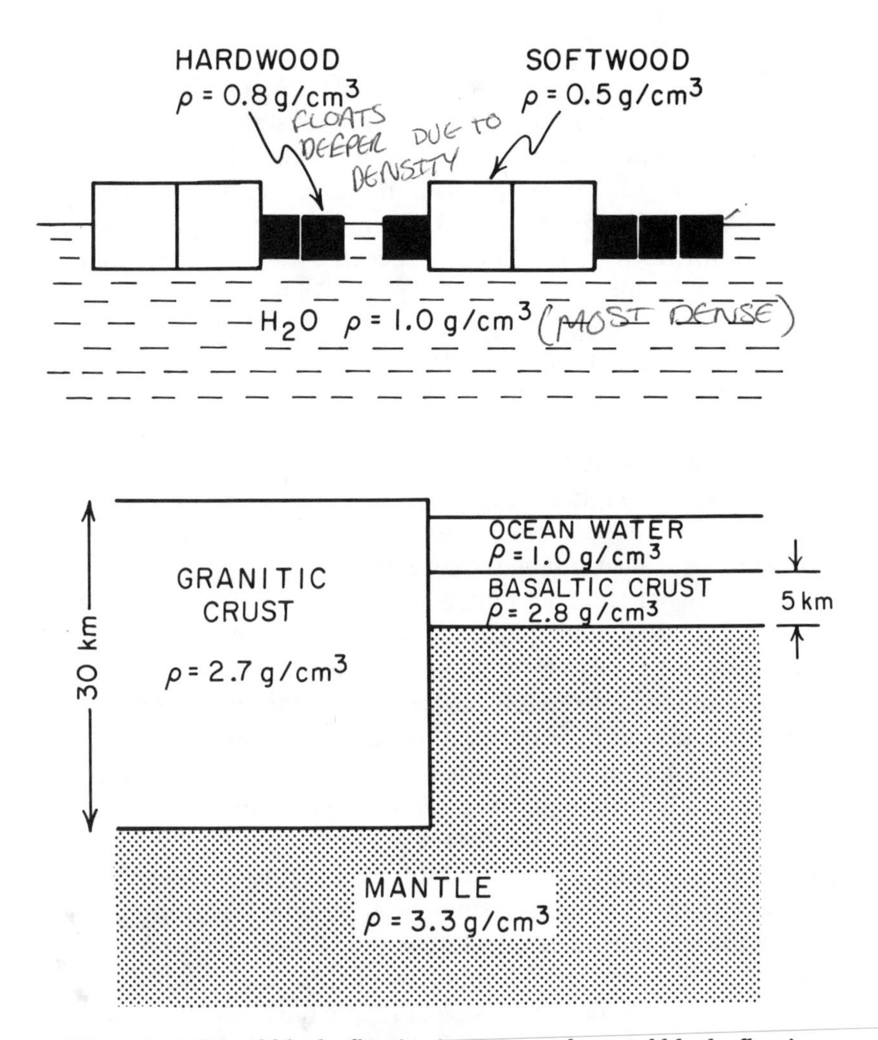

Figure 5-11. Wood blocks floating in water and crustal blocks floating in the mantle: The upper panel shows the relative heights of two kinds of wood blocks floating on water. The softwood stands higher partly because the blocks are thicker and partly because they are less dense. The lower panel shows the relative heights of granitic and basaltic crust floating on the mantle. The granite blocks stand higher because they are thicker and less dense. Because the basaltic blocks stand lower, the Earth's oceans are located above them. The weight of this water pushes down the basaltic blocks a bit further. The symbol ρ used in the diagram stands for density.

Age of the Crust

Now what about the ages of the crusts? Again the situation for the ocean floor is quite simple. The crust starts at the oceanic spreading centers with an age of zero. It takes somewhere between 50 and 150 million years for this newly born crust to move across the sea floor to a zone of convergence where it disappears back into the mantle. Thus the oceanic crust is very young because it is continually being renewed. Each complete renewal takes, on the average, only about 0.1 billion years. If the mantle cells which cause the oceanic crust to be renewed have run at the same rate during all of Earth history, then the Earth has had 45 successive oceanic crusts! The present distribution of ocean crust ages is shown in Figure 5-12.

As shown in Figure 5-13, the geographic pattern of ages for continental material is not nearly so regular as that for the oceans. Also, the age range is huge, from 0.1 to 3.8 billion years! No age greater than 3.8 billion years has been found. If older rocks exist they must constitute only a very small part of the crust, since the search for such rocks has been intense.

The limiting ingredient for the production of granite is the element potassium (K). As shown in Table 5-3, granite has about 160 times more K than mantle material. Today about a half of the Earth's K is thought to reside in the continents. Although no one knows for sure the details of the route taken by K from mantle to continent, one possibility is that it is via ocean-floor basalts. If so, we can place some limit on how fast granites are being made. Oceanic basalts contain only 0.1 percent K. Average continental crust contains about 1.0 percent K. Thus, about 10 units of oceanic crust are needed to supply the K needed to make 1 unit of continental crust. This means that taken together *all* 45 of the successive oceanic crusts could produce an amount of continental crust no greater than 4.5 times the volume of the present oceanic crust (45/10 = 4.5). This volume can be compared with the actual volume of continental crust. The area of the continents is about one-half that of the ocean basins. The average thickness of the continental crust (30 km) is about 6 times greater than the average thickness of the oceanic crust

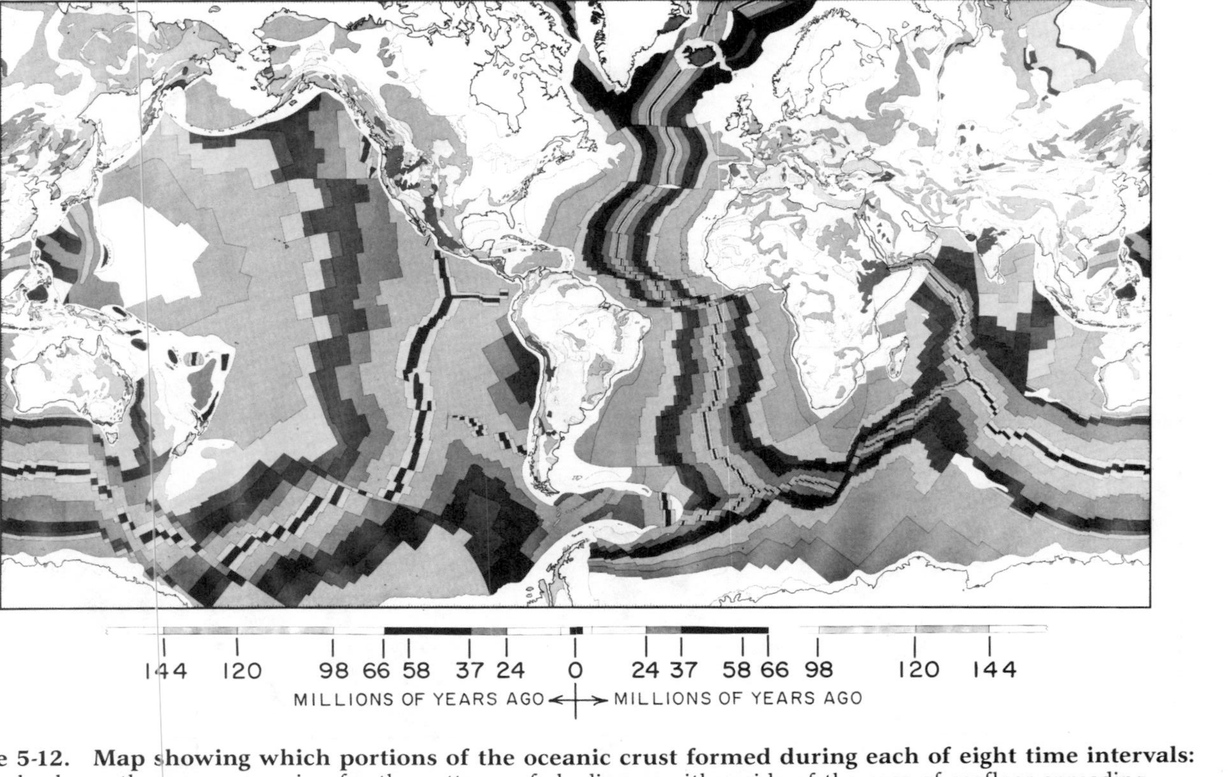

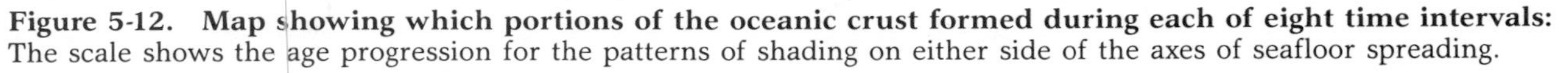

Figure 5-12. Map showing which portions of the oceanic crust formed during each of eight time intervals: The scale shows the age progression for the patterns of shading on either side of the axes of seafloor spreading.

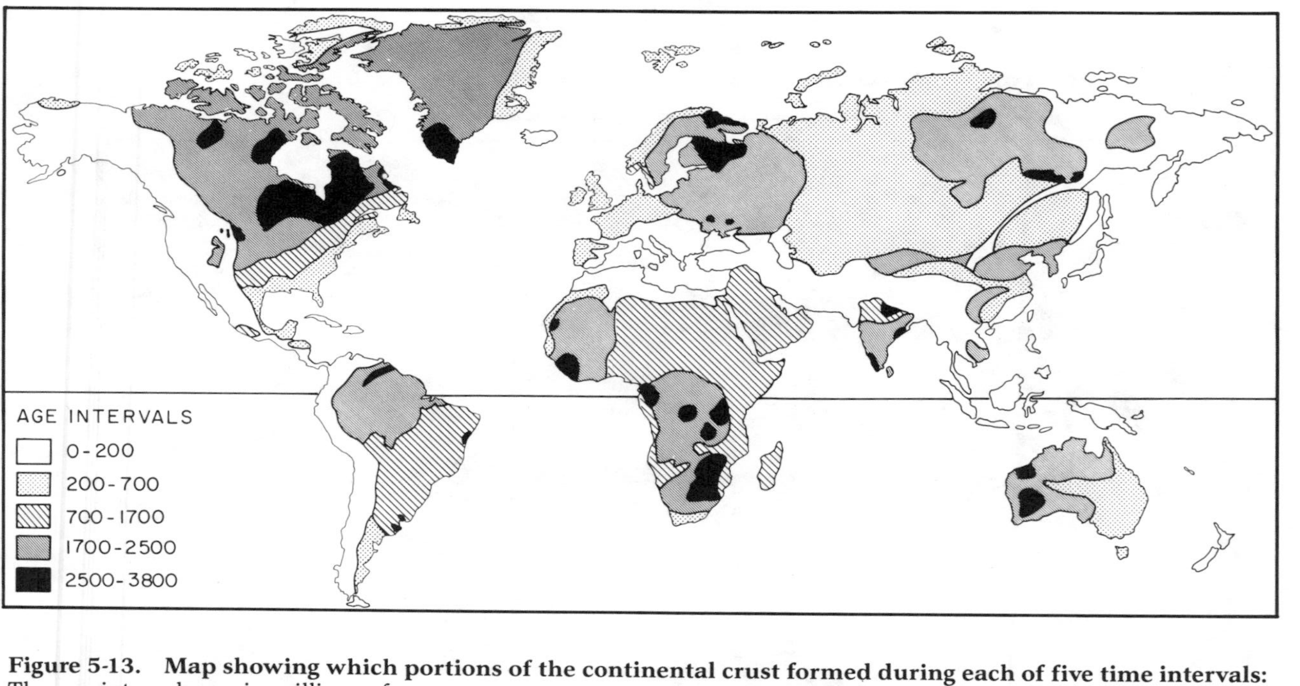

Figure 5-13. Map showing which portions of the continental crust formed during each of five time intervals: The age intervals are in millions of years.

(5 km). Thus, the volume of the continental crust is 3.0 times (0.5 × 6) that of the oceanic crust. Since some of the continental crust that formed in the past must have been subducted back into the mantle, the current amount of crust is smaller than the amount manufactured. While this calculation is certainly not a very accurate or even a very realistic representation of what has happened, it does tell us a very important thing. If we use the current rate of delivery of K from the mantle to the oceanic crust as a guide, several billions of years are required to build continents!

Summary

The evidence in hand suggests that the Earth's core and atmosphere formed during the first 2 percent of its history. By contrast, the buildup of the continents appears to have begun much later and continues today.

The continental crust has a chemistry far different from that of the planet as a whole. Although rich in silicon like the mantle, it has far higher iron and magnesium contents and far lower aluminum, sodium, and potassium contents.

The continents are buffeted about by the convecting mantle. This buffeting has thickened them to the point where erosional thinning keeps pace.

Supplementary Readings

The Inaccessible Earth, *by G.C. Brown and A.E. Mussett, 1981, George Allen and Unwin.*

A readable treatment of the structure and internal composition of the Earth.

Inside the Earth, *by Bruce A. Bolt, 1982, W.H. Freeman and Company.*

An introduction to the use of seismic waves to probe the Earth's interior.

The New View of the Earth: Moving Continents and Moving Oceans , *by Seiya Uyeda, 1978, W.H. Freeman and Company.*

An introduction to plate tectonics.

Marine Geology, *by Roger N. Anderson, 1986, John Wiley and Sons.*

An up-to-date treatment of the physics and chemistry of the ocean lithosphere.

Comet Arend-Roland

Chapter Six

Contending with the Ne

Moons, Asteroids, and Co...

Not all the material in the solar system resides in the Sun and its nine planets. Six of the planets are orbited by moons. A belt of 10 billion asteroids lies between the orbits of Mars and Jupiter. Myriad comets whiz around the Sun in orbits far larger than that of Pluto.

Despite the great treasure of information obtained from our manned voyages to the Moon, the origin of this object remains a mystery. While showing a strong isotopic affinity to Earth matter, elements of moderate volatility are highly depleted in the Moon compared to Earth. To account for this difference it appears to be necessary that the matter now in the Moon once orbited about the Earth as a ring of very hot debris.

The asteriods and comets are thought to be remnants of the small objects that dominated the solar nebula at the onset of planetary growth. Because their orbits were just right, they have thus far avoided falling into the gravitational grip of one of the neighboring planets. The meteorites that fall to Earth bear witness, however, that the big objects are still gobbling up the small-fry. While most meteorites are too small to do serious damage, we have evidence that now and then objects many kilometers in diameter hit the Earth. One such collision 65 million years ago caused half of the species of living organisms to become extinct!

Introduction

In addition to the Sun and the nine planets, the solar system contains a host of other objects. The most prominent of these are the 33 moons that circle the planets. The list does not end here. Circling the Sun between the orbit of Mars and the orbit of Jupiter are about 100 billion asteroids. While most of them are 1 meter or less in diameter, about 2000 are greater than 10 kilometers in diameter. Only one, Ceres, is more than 1000 kilometers in diameter. Jupiter and Saturn have rings made up of myriads of small objects. Finally, on the order of 1000 billion comets* are thought to ride in orbits more distant than Pluto's.

We have two major interests in these objects. The first is a hope that they carry important information about the origin of the solar system. The other is that by colliding with the Earth the odd fragments of solar-system matter appear to have influenced the development of life.

Moons

Six of the Sun's nine planets have their own satellites. We call these objects moons. As shown in Figure 6-1, all four major planets have moons. Jupiter has 13, two which are larger than the planet Mercury! Saturn has ten, one of which is larger than the planet Mercury! Uranus has five and Neptune has two. Earth and Mars are the only terrestrial planets with moons. We have one, Mars has two.

The space probes sent by NASA to explore the solar system have photographed the moons of Mars, Jupiter, Saturn, and Uranus, showing them to be solid objects dotted with craters made by the impacts of meteorites (see Figures 6-2 and 6-3). Because of the cold environment in which the major planets formed, their moons are likely to be considerably more rich in

*While we have no direct chemical information, scientists believe that comets consist of ices wrapped around a rocky core. In this sense, comets are to the major planets what asteroids are to the minor planets, namely, miniature versions. The famous "tails" of comets are thought to be the vapor created when in the course of their highly elliptical orbits they come close to the hot Sun. Scientists recently sent out several space probes which telemetered back information about Halley's Comet as it passed close to the Sun in 1986.

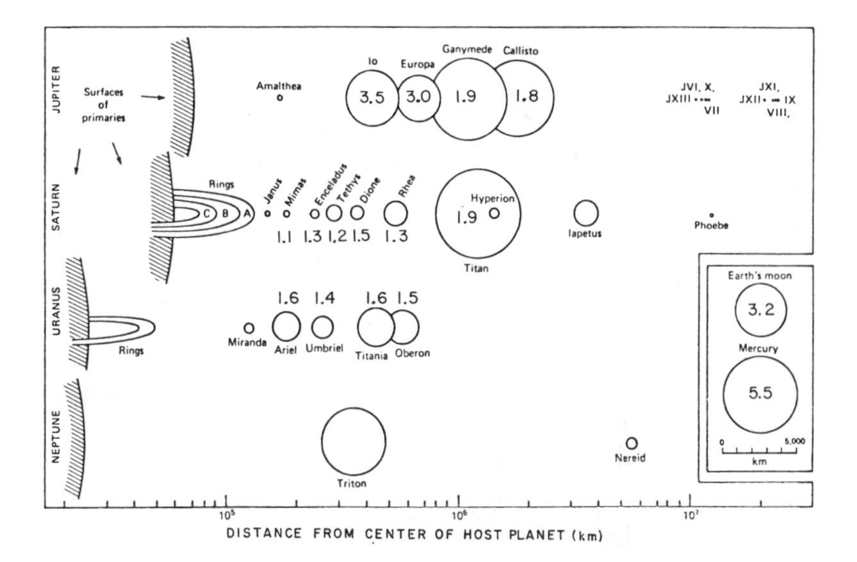

Figure 6-1. Sizes of the moons of the four major planets: As can be seen, three are larger than Mercury and five are larger than the Earth's Moon. Also shown are the spacings of these moons relative to the center of the host planet (the left-hand edge of the diagram) and of the planet's surface and rings. Note that the distance scale is logarithmic. The numbers give the bulk density of the object in (gm/cm^3). While Io and Europa have densities close to that of our Moon, the other moons have much lower densities and hence must contain a substantial component of the ices. No masses are given for the as yet unexplored planet Neptune.

volatile elements than are the terrestrial planets. Indeed, as indicated by the densities given in Figure 6-1, some of these moons cannot consist entirely of magnesium, silicon, iron, and oxygen. Rather, they must contain substantial amounts of the ices. Too little is known to date about these objects for us to warrant further discussion. We will concentrate instead on the much richer information available about our own Moon. One point must be stressed, however: moon formation is not a rarity. There are 33 moons and only nine planets! Hence, the formation of these objects must be as natural a consequence of the processes that produced our solar system as is the formation of planets.

The Earth's Moon

Even before our astronauts landed on the Moon, we were quite sure that it had no air, no ocean, no ore deposits, no oil, and no life. The observations made by our astronauts while on the Moon and studies of the rocks they brought back were aimed toward unraveling the mystery of the Moon's origin. Rather than revealing the secret of its origin, the Moon has shown itself to be even more of a maverick than we thought. As we shall see, both the Moon's orbit and its bulk chemistry are difficult to explain. Taken together, they pose one of the most perplexing puzzles in planetary science.

Perhaps the best way to introduce this puzzle is to list the four scenarios which have been posed for the Moon's origin. They are shown by the cartoons in Figure 6-4:

1) *The fission hypothesis*. One possible way to create the Moon is to have the earth spinning so fast early in its history that it cast out a material from its mantle. This material became the Moon.
2) *The capture hypothesis*. Another possibility is that the Moon formed elsewhere in the solar system and by chance was subsequently captured by the Earth's gravitation.
3) *The accretion hypothesis*. Still another possibility is that the Earth and Moon formed side by side from the same pool of nebular material.
4) *The collision hypothesis*. A more recent proposal is that the material making up the Moon was ejected from Earth when it was struck by an object the size of the planet Mars.

Before discussing the relative merits of these hypotheses, let us consider two key pieces of information critical to distinguishing among them.

The Moon's Chemical Composition

As outlined in Chapter Three, the bulk density of any object tells us something about its chemical composition. We saw that the range of densities observed for the terrestrial planets could be matched by simply changing the proportions of the four dominant elements (oxygen, magnesium, silicon, and iron). The

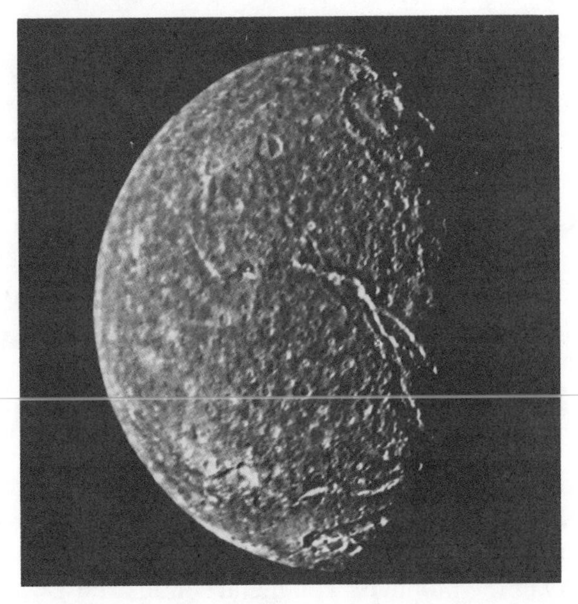

Figure 6-2. Photographs of the surfaces of Saturn's moon, Rhea (top) and Uranus's moon, Titania (above).

Figure 6-3. Photographs of the surface of the Martian moon, Phobos.

moon lies at the low end of the planetary density spectrum. It has a bulk density of 3.2 gm/cm³. This is quite a bit lower than the value of 4.3 gm/cm³ for the Earth. To explain the low density of the Moon, its iron metal content must be much lower than that for Earth.

Also, Moon rocks have a distinctly lower potassium to ura-

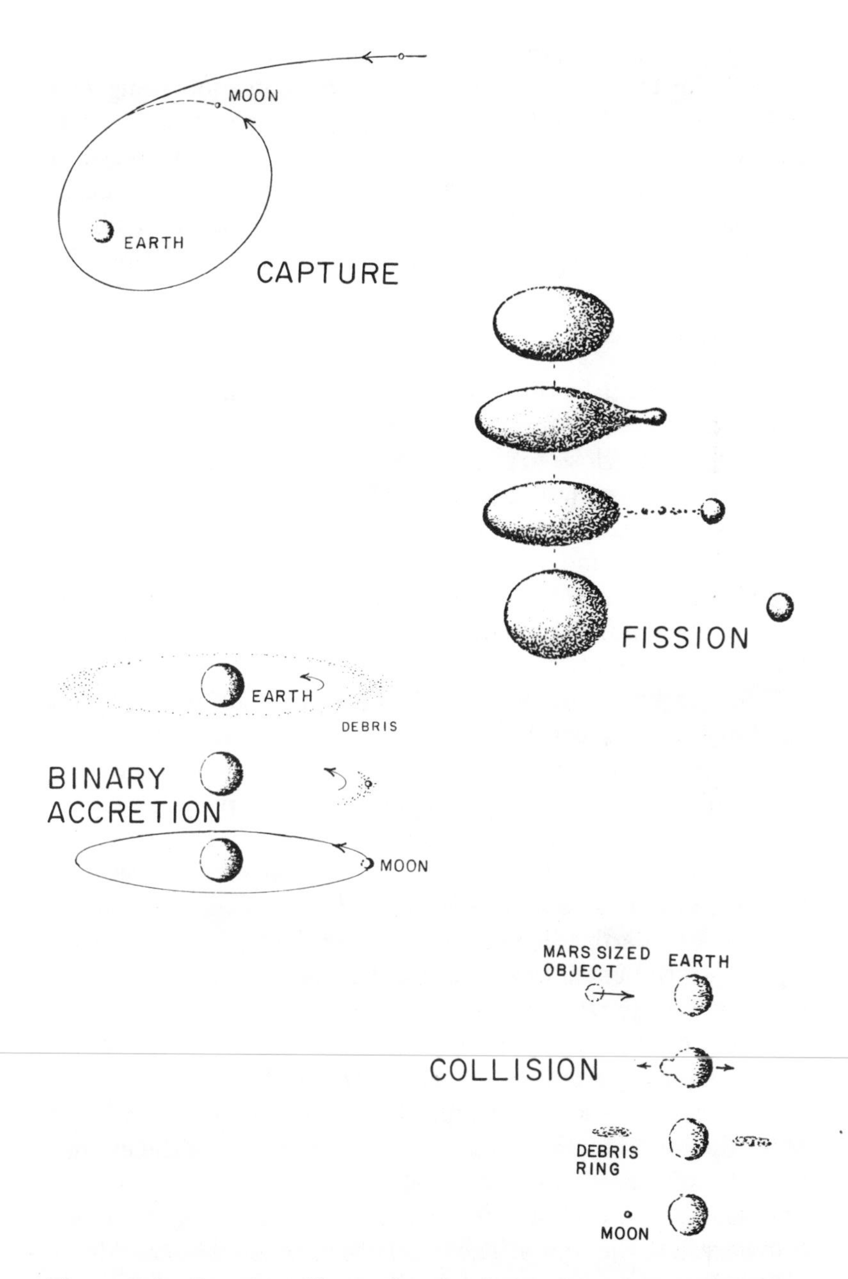

Figure 6-4. Drawings illustrating the four origins hypothesized for the Moon.

nium ratio than Earth rocks. For the Earth the K/U ratio is 12,000. For the Moon it is only 2000. This difference suggests that the material that formed the Moon was subjected to temperatures considerably higher than was the case for the material that formed the Earth. Estimates based on the studies of Earth and Moon rocks suggest that the Moon is also highly depleted in all elements more volatile than potassium (see Figure 6-5).

The Moon's Orbit

The important features of the Moon's orbit are:
1) it is almost perfectly circular;
2) it is quite close to the Earth (see Figure 6-6);
3) it is growing larger with time.

As we shall see, the last of these three makes the first two even more difficult to understand. If the Moon has been with us from the beginning, then it must have started in a very close orbit. Hence, we will first consider the evidence for and the cause of the Moon's retreat.

While we have long suspected that the Moon is retreating from the Earth, only recently has dramatic proof appeared. One of the chores given the astronauts who visited the Moon was to put in place so-called corner reflectors. Their purpose was to provide precise points from which a laser beam shot from the Earth could be bounced back to Earth. By measuring precisely the time required for a pulse of laser light to travel to the Moon and back to Earth it is possible to establish the distance from a point on the Earth to a point on the Moon with an accuracy of about 1 centimeter! These measurements have been repeated regularly over a period of ten years. They show that the Moon is moving away from the Earth at the rate of 4 cm per year.

Scientists have long recognized that the energy required to "lift" the Moon further from the Earth comes from the energy bound in the Earth's spin. Through a process called tidal friction, energy associated with the Earth's spin is being gradually transferred to the Moon. The extra energy gained by the Moon speeds its movement around the Earth and lifts it to ever more distant orbits.

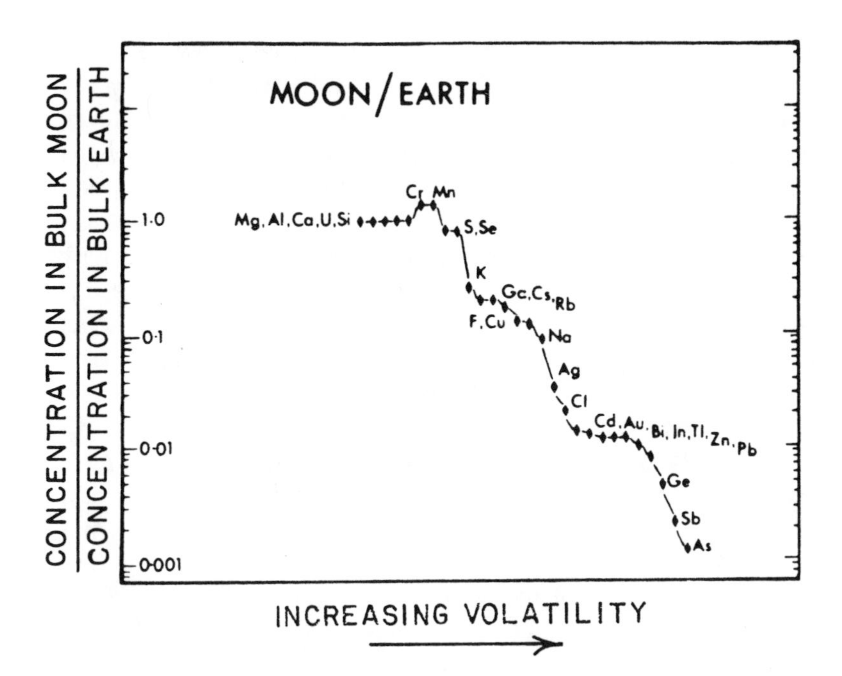

Figure 6-5. Depletion factors for volatile elements in the bulk Moon relative to the bulk Earth.

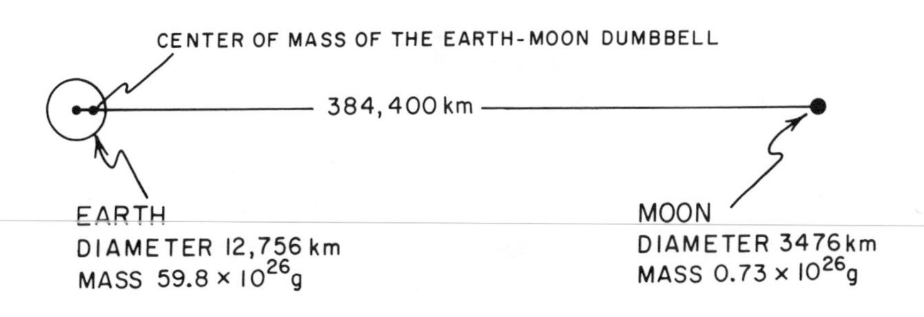

Figure 6-6. Properties of the Earth—Moon system: The Moon follows a nearly circular path about the Earth. This orbit is located 30 Earth diameters away from us. The pull on the Moon's mass (1.2 percent that of the Earth) causes the Earth to wobble in its orbit about the Sun. It is the center of mass of the Earth—Moon dumbbell that follows a smooth course about the Sun.

Anyone who has spent time at the seashore knows that the tides rise and fall twice a day. These tides are driven by the gravitational attraction exerted by both the Sun and the Moon. Tides also have a monthly cycle related to the position of the Moon relative to the Sun. At the new Moon, the Sun and Moon tug together. At full Moon, the Sun pulls one way and the Moon the other. The Earth's water bulges are never quite where gravity would like them to be; rather, they are always on the move trying to keep pace with the Earth's spin. Instead of pointing in the direction demanded by the pull of the Sun and Moon, the Earth's tidal bulges point in the direction demanded some hours earlier. The Earth's spin keeps carrying the bulges forward of the position demanded by the pull of the Moon and Sun. The water eternally rushes back toward the required positions. Since it can't do so quite fast enough, the water bulges are always somewhat out of place.

The Moon does not have a perfectly spherical shape. These irregularities give it gravitational bulges. The Moon's bulge has become "locked" to the Earth. This is why we never see what we call "the back side" of the Moon. The Moon's day is one month long! Because of this gravitational locking, the Moon's bulge is always right where gravity wants it to be—pointing directly at the Earth.

The Earth's water bulge exerts a gravitational pull on the Moon. This pull speeds the Moon's movement around the earth. Correspondingly, it retards the Earth's spin. So tidal friction causes the Moon to speed up and the Earth to slow down.

The Moon is currently 3.8×10^{10} centimeters from the Earth. If the retreat rate was always 4 cm per year, then the Earth and the Moon would have been touching about 9 billion years ago. However, the closer the two objects are to each other, the stronger the tidal friction. Hence, the retreat has slowed down as the Earth–Moon distance has increased. Other things being equal, when this is taken into account the time for touching comes out to be only about 2 billion years. As the Earth formed 4.6 billion years ago this estimate of the "touching age" is embarrassingly small. This figure cannot be taken too seriously, however, since the geography surely changed with time. Unfortunately, there is no way to reconstruct the sizes, positions, and depths of the Earth's seas in the distant past. Thus, all we can say is that the rate of retreat is such that the Moon must have been much closer to us early in Earth history.

About two decades ago, John Wells, a paleontologist at Cornell University, came upon a potential means for determining the extent of the Moon's retreat on a much longer time scale than that covered by NASA's measurements. Wells was aware that corals living in today's reefs are banded. The most prominent of these bands have been shown be be annual. The porosity of the calcium carbonate deposited by these organisms changes subtly with the seasons. These changes can be seen on medical x-rays taken of slabs cut from a coral head (see Figure 6-7). In addition to seasonal bands, Wells also saw weaker banding which he attributed to the monthly tidal cycle and to day–night cycles.

If the Moon has retreated at a rate of 4 or so centimeters per year, then several hundred million years ago there must have been both more days and more months in a year. While the day part is clear, the month part is not. To understand it, we must refer back to Kepler's Law (see Chapter Three), which states that the further a satellite orbits from its gravitational master, the longer it takes the satellite to complete its round. While the Moon is being made to move faster by the tidal tug of the Earth, the circumference of the Moon's orbit also increases. It turns out that according to Kepler's Law the fractional increase in the circumference of the orbit is larger than the fractional increase in the speed at which the Moon moves! For example, a 20 percent change in circumference leads to a 31 percent change in period. Thus, despite the fact that it moves ever faster, the Moon takes ever longer to complete each revolution (the current time is 27.3 days). So the length of the "month" increases as the Moon retreats. The Moon was able to make more trips around the Earth in one year in the distant past than it does now.

Wells found that in fossil corals about 0.36 billion years old the number of daily bands associated with each annual layer

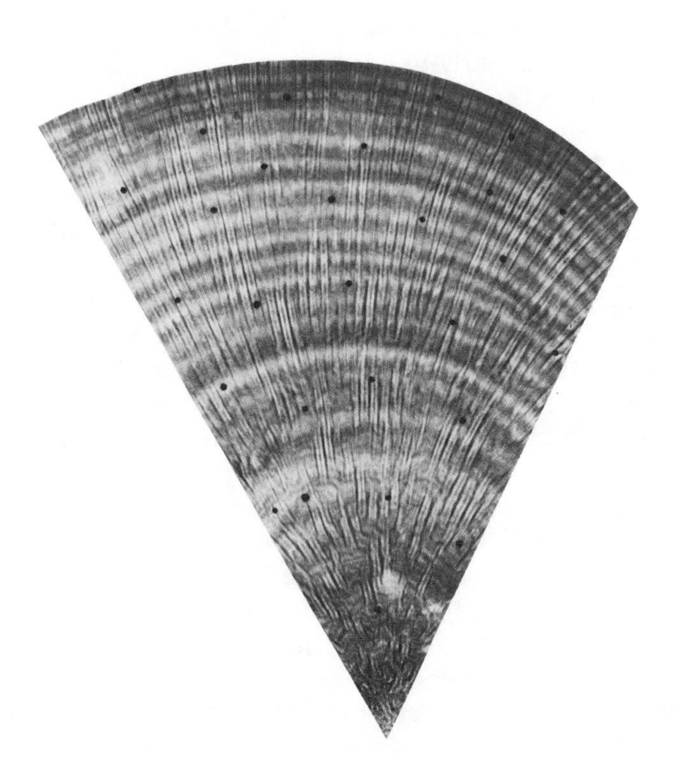

Figure 6-7. X-ray photograph of a slab of coral: The annual growth rings are clearly visible. The dark bands represent growth during summer months. Proof that the most prominent growth bands in coral heads are caused by the seasonal changes in the character of the calcium carbonate they deposit was provided by studies of corals from the Eniwetok Atoll. The test of an early version of the hydrogen bomb conducted in this atoll in 1954 produced an enormous amount of local radioactive fallout, so much that the waters of the atoll became temporarily highly contaminated with fission fragment ^{90}Sr. Since the element strontium substitutes readily for the element calcium in the $CaCO_3$ formed by coral, the 1954 growth band was "marked" with radioactive strontium. When corals collected a decade or more after this event were analyzed it was found that there was one growth band for each year that had elapsed since the 1954 test. Thus the ^{90}Sr marking of the 1954 band allowed the annual growth-band hypothesis to be verified.

was different than expected! He found about 400 daily bands. As there is no reason to suspect that the time required for the Earth to orbit the Sun has changed significantly, these results suggest that the day was shorter in the past than it is today. The changes found by Wells in the number of days tell us that the Moon was 1.2×10^9 cm closer to the Earth 0.36 billion years ago than it is today. Thus, if the record in fossil corals is being properly read, then the Moon's retreat rate over the last 7 percent of Earth history has averaged 4 centimeters per year, just as it has over the last ten years.

Which Hypothesis for the Moon's Origin Is Correct

It was the retreat of the Moon that hooked Darwin and other early students of the Moon on the fission theory. If one calculates the length of the day and of the month back to the point where the Earth and Moon were together, both come out to be about six hours. This higher spin rate supposedly allowed the Earth to fission. This calculation is independent of the actual rate of retreat; it depends only on the current combined angular momentum of the Earth and Moon.

It is generally agreed that in order to fission, the Earth would have to have been spinning at a frequency equal to its critical vibration frequency. The situation is similar to that which occurs when an object is subjected to sound having a frequency matching the natural vibrational frequency of the object. In this situation the amplitude of the object's vibrations becomes very large. The well-known example is that of the opera singer who causes a wine glass to shatter by adjusting the pitch of her voice to match the natural frequency of the wine glass. The same physics could cause a spinning body to fission. The Earth may have been torn apart by the "sound" of its own spin.

What Darwin didn't know was that the Earth's natural frequency was two hours rather than six hours. Thus, with its six-hour day, the combined Moon–Earth system would not have spun fast enough to induce the opera singer effect. Another consideration that Darwin missed was that the Moon cannot be a discrete blob thrown from the spinning Earth, for any such blob would have been torn to pieces by the Earth's gravity as it pulled clear of the Earth. Rather, if fission is the correct scenario, the Moon must have accumulated from the ring of debris generated by the fission event.

The only way to circumvent these dynamic problems is to assume that a mass of material several times the Earth's present mass was thrown from the Earth at the time of its fission. If so, then the Earth could have been spinning fast enough to undergo a wine-glass type of disruption. Since the lion's share of this required extra material is no longer with us, it must have escaped the Earth–Moon gravitational field. It could have done so only if it was a gas. To have been in gaseous form, the debris flung from the Earth must have been immensely hot. This could explain the Moon's deficiency in volatile elements.

If it is admitted that the blob fissioned from the Earth shattered into a ring of hot debris, then from this point on the fission hypothesis becomes similar to the remaining three hypotheses. There is an important difference, however. It lies in the chemistry. Had the Moon originated by fission, then there is a ready explanation for its low iron content. Mantle rather than core material was spun off. Of course, for this to be the case the Moon could not have formed until after the Earth's core formed.

Despite all their efforts, geophysicists have not found a reasonable route by which the Moon might have been captured by the Earth. Successful scenarios all require a highly improbable solar orbit for the Moon before its capture and a very lucky "catch" by the Earth. As scientists shun explanations that require "one chance in a billion" events, this explanation currently receives little enthusiasm.

The more probable routes of capture all lead to near encounters with the Earth during which the captive is torn asunder. The debris from this disruption would then have to be reassembled to form the Moon. At this point the capture theory joins ranks with the fission scenario.

For the Earth and the Moon to form from the same pool of

material and yet end up with quite different compositions requires that the iron in the nebula was accumulated preferentially in the Earth. For some reason, a small portion of the debris which might have become part of the Earth's mantle remained in a ring orbiting the Earth. At this point this scenario merges with the first two.

Faced with the lukewarm reception received by the three long-standing scenarios, scientists have come up with a fourth. According to this, the material for the Moon is splashed from the Earth during the impact of a Mars-sized object. The ultrahot matter ejected as the result of this impact forms a belt of debris which eventually cools and aggregates to form the Moon. To avoid the awkward improbability that a tenth planet by remote chance collided with Earth (a scientific no-no), it is proposed that early in the history of the solar system much of the solar system's matter was in the form of small planets. These eventually were gobbled up by collisions with larger objects. One of these was a late impact on Earth that blasted out the material now found in the Moon.

To be successful, it appears that all four scenarios require that the Moon formed from a ring of debris which orbited Earth. Were the debris to have been very hot, then each scenario has a means to deplete the Moon in volatile elements. Thus, the Moon appears to be telling us that it originated from a debris ring. We have no good clues regarding how this ring came into being, however. Was it created by disruption of a blob of material fissioned from the Earth? Was it created by the disruption of an object captured by the Earth's gravitational field? Was it a remnant of the material from which the Earth formed? Was it blasted out during the collision of a Mars-sized object? We don't know.

Oxygen-Isotope Evidence

There is one more piece of evidence bearing on the Moon's origin. It is isotopic. As shown in Figure 6-8, the oxygen-isotope composition of Earth rocks differs from that for the two major classes of chondritic meteorites. Earth rocks and minerals show a spread in their oxygen-isotope compositions falling along a

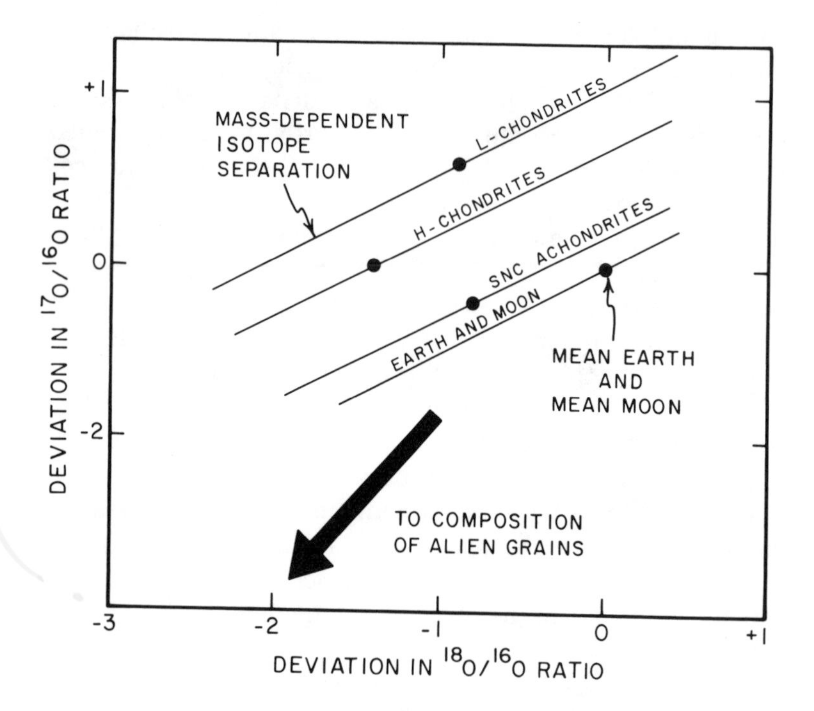

Figure 6-8. Oxygen isotope composition for Earth, Moon, and meteorite rocks and minerals: The slopes of these trends are those expected for mass-dependent fractionation taking place during the melting and crystallization. The displacement of the line for the Earth from those for the two groups of meteorites presumably stems from differences in the extent to which the debris from which these objects formed was contaminated with alien grains. By contrast, Moon and Earth materials have isotopic compositions lying along the same line! Based on their chemical composition and age, the SNC achondrites are thought to be pieces of Mars blasted off during impacts. If so, then Mars has an oxygen-isotope composition different from the asteroids and the Earth–Moon system.

mass fractionation line. Analyses for meteorites also fall along three such lines. However, the meteorite lines differ from the Earth line (and from each other). One line is for minerals from high-iron meteorites, another is for minerals from low-iron meteorites, and the third is for a class of basaltic achondrites. The most likely explanation for the displacement of these lines from one another is different extents of incorporation of alien grains (see Chapter Four). Regardless of the explanation, because of the differences in their oxygen–isotope composition it seems logical to conclude that these three types of material formed at separate places in the solar nebula.

The important point to be made is that no difference exists in this regard between Moon rocks and Earth rocks. The oxygen–isotope composition for rocks from both objects lies along the same mass fractionation line! This concordance is what would be expected if the material in the Moon was derived by fission. It is also to be expected if the Earth and Moon accreted side by side. On the other hand, had the Moon formed elsewhere and been captured, one might expect that like meteorites its oxygen-isotope composition trend line for lunar material would be different from that for Earth material. Thus, the oxygen evidence gives us a reason to eliminate the capture hypothesis from contention.

Layering of the Moon

Instruments left on the Moon by our astronauts have radioed back seismograms generated by moonquakes. Although not conclusive, these results suggest that the Moon has a small core (about 2 percent of its mass). This is consistent with the Moon's low bulk density.

The Moon has no atmosphere or ocean. The reason is that, regardless of mass, gaseous molecules can readily escape from the Moon's surface. Here we encounter one of the prerequisites for habitability. If a planet is too small, it can retain neither an atmosphere nor an ocean. (More will be said about this in the next chapter.)

Studies of lunar rocks reveal that the Moon must have a crust. The Moon's crust can be divided into two parts, the maria and the highlands (see Figures 6-9 and 6-10). The Moon's maria are made of a rock akin to the Earth's basalt and its highlands of rocks somewhat akin to the Earth's granite; the highlands rock, called anorthosite, consists primarily of the mineral calcium feldspar: $Ca\ Al_2\ Si_2\ O_8$. Here the similarity between the Earth's crust and the Moon's crust ends. Dating of the Moon's basalts shows that they all formed mainly during the period 3.1 to 3.9 billion years ago. The anorthosite crust is even older, having formed more than 4.0 billion years ago. Little or no volcanism has occurred on the Moon during the last 3 billion years. Since then the Moon has been a "dead" planet. No convection cells churn in its mantle. No plates collide on its surface. No volcanoes erupt.

Why this big difference? Again, it is the Moon's small size that is responsible. The heat released by the decay of the Moon's long-lived radioisotopes can more easily get out. Great convection cells are no longer needed to carry heat from its interior to the top of its mantle.

Students of the Moon see its history as follows. Sometime before 4 billion years ago the Moon melted and then slowly solidified. This solidification began with the crystallization of calcium feldspar crystals. Being less dense than the residual liquid, these crystals floated to the surface and formed a thin crust. During the period 4.1 to 3.9 billion years ago this crust was punctured by a series of large meteorite impacts. Over the next billion years basaltic magmas welled up into these craters, creating the maria. As can be seen from Figures 6-9 and 6-10, the maria are all located on the side of the Moon that we see. The back side has an uninterrupted highland terrain.

Impact Craters

Photographs of the Moon (Figures 6-9 and 6-10), of Mars (Figure 6-11), of Mercury (Figure 6-12), and of the moons of Jupiter, Saturn, and Mars (Figures 6-2 and 6-3) show surfaces pockmarked by the craters made by the impacts of meteorites. By contrast, the Earth's surface is nearly free of these blemishes. True, the Arizona meteor crater (Figure 6-13) is to be seen. Were one to examine photographs of the Earth taken from space, however, it would be a long while before another such example could be found. Undoubtedly the Earth has received its share of these planetary oddbits. Its relative smoothness is a tribute to the erosional processes that continually modify the Earth's surface, obscuring and ultimately erasing impact features.

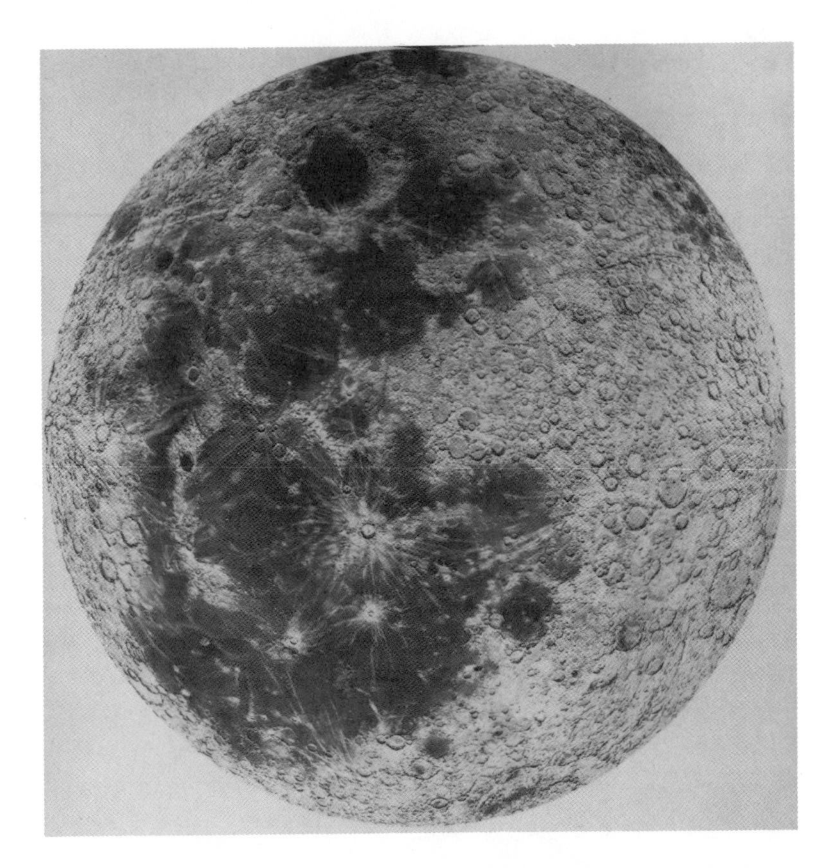

Figure 6-9. The Moon's front side: The dark-colored portions (called maria) are areas of the Moon surface that were flooded by lunar basalt flows. The light-colored portions (called highlands) represent the original anorthositic crust; these show the pockmarks produced by countless meteorite impacts.

Although the Earth's surface is largely free of the visual imprint of impacts, clear evidence for their occurrence is to be found. One such source of evidence comes from glassy objects called tectites (see Figure 6-14). Based on their aerodynamic shapes and surface textures, tectites are thought to have been formed from liquids which crystallized to glass above the Earth's atmosphere and then were modified by ablation as they fell back

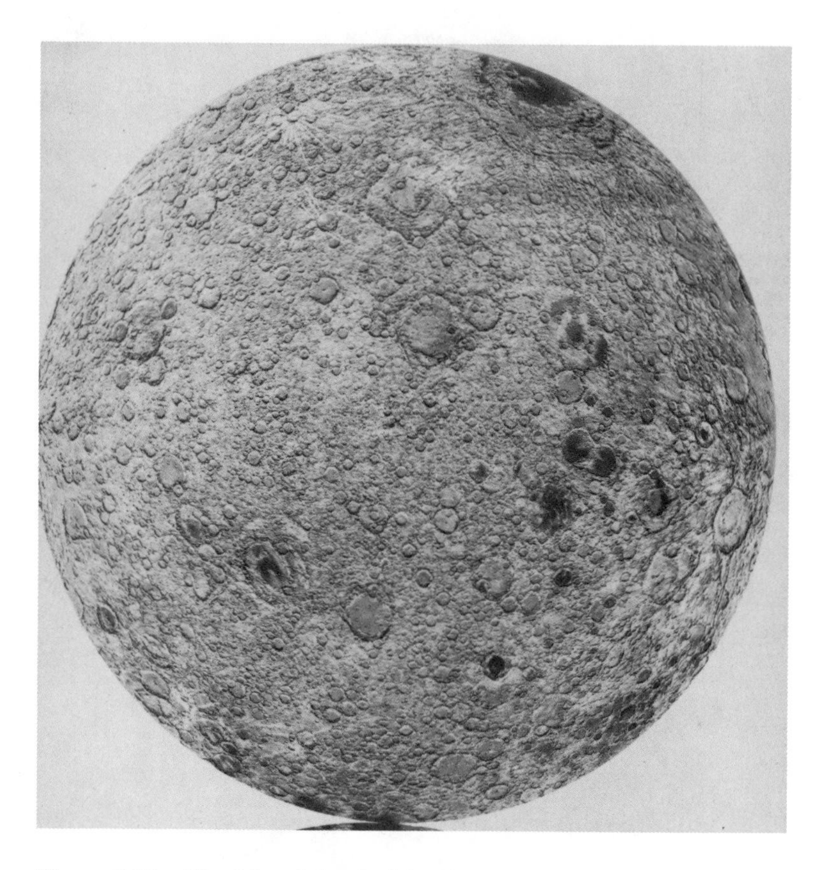

Figure 6-10. The Moon's back side: Since this part of the surface never faces the Earth, it has only been seen by the astronauts. Note that the maria, which are so prominent on the Moon's front side, are nearly absent on the back side.

through the atmosphere to the Earth's surface. Most scientists are convinced that these objects were formed during impacts which splashed material up from the Earth's surface well above the atmosphere.

The most abundant of the tectites are those found throughout Southeast Asia and Australia in soils and streambeds and in marine sediments adjacent to these land masses. These tectites have ages of 0.7 million years and are thought to have been

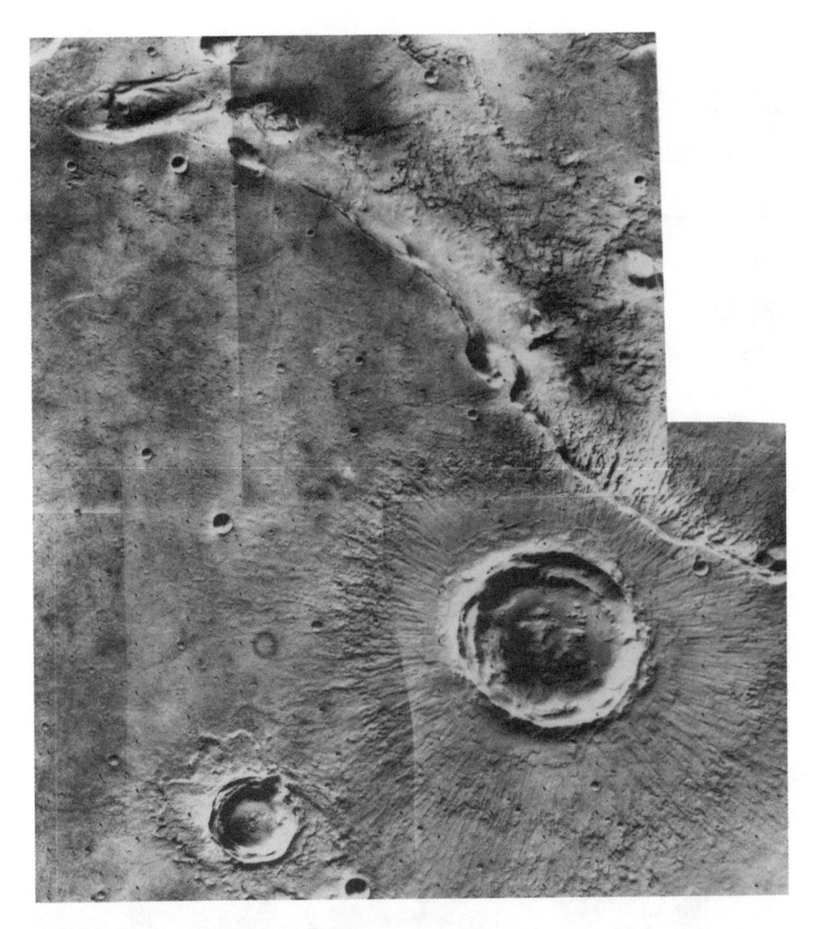

Figure 6-11. The surface of Mars: A landscape dotted with impact craters.

formed by a single large impact. Another group of tectites is found in North America (age 30 million years), another in central Europe (age 13 million years), and another in the African Ivory Coast (age 1.1 million years).

In the case of the European tectites, the actual impact crater is thought to have been located in Germany. Even though it has been partially erased by erosion over the last 30 million years, enough remains to prove its origin. For example, sedimentary rocks deformed by this impact contain SiO_2 minerals formable

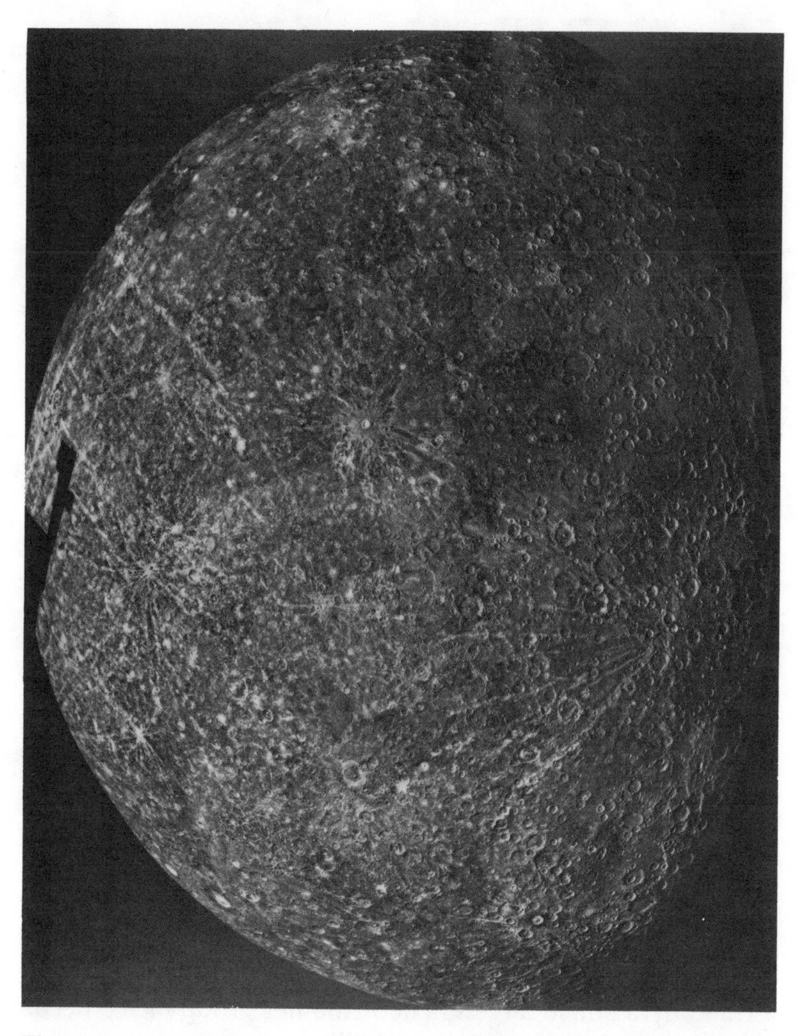

Figure 6-12. The surface of Mercury: A landscape dominated by impact craters.

only at high pressure. These minerals were formed from ordinary quartz by the great pressures associated with a high-speed impact.

One might ask what would happen were the Earth to be hit by an oddbit big enough to make one of the larger of the craters

Figure 6-13. The Arizona meteor crater in the desert of the southwestern United States: It is one of few impact pockmarks seen on the Earth.

Figure 6-14. Tectites: Black glass objects whose shapes and textures bear witness to high-speed flight through the atmosphere. They are thought to have been created from material melted and splashed high above the atmosphere by the impact of meteorites or comets.

we see on the surface of other planets and moons. Surprisingly enough, we think we know of at least one such event. It happened rather late in Earth history (about 0.065 billion years ago). While geologists have suspected for many years that some sort of catastrophe must have occurred at this time,* only recently has evidence appeared linking this event to the impact of an asteroid (or of a comet). What they did know was that about one-half of the species of plants and animals present during late Mesozoic time suddenly became extinct. A wide variety of hypotheses have been proposed to explain this massive blow to life, but since none of these could be supported by any hard evidence, the cause of this crisis remained a mystery.

Then in 1980 Luis Alvarez, of the University of California, Berkeley, and his son Walter undertook a study which broke the problem open. Walter Alvarez had been studying a continuous sedimentary sequence in Northern Italy encompassing the boundary between the Mesozoic and Cenozoic eras. He collected a series of sediment samples which began well below the boundary and ended well above it. His father, working with Frank Asaro and Helen Michel, had these samples analyzed for their content of the rare element iridium. To their great pleasure and amazement, the Alvarez team found that the sample taken right at the boundary had 60 times more iridium than samples from above and below the boundary. They proposed that the explanation for this peak in iridium was that a 10-kilometer-diameter asteroid had hit the earth and blown to bits. Debris from this explosion was carried high above the atmosphere and distributed around the Earth. Within a few years this debris settled back reaching every part of the Earth's surface. That which fell on the surface of the sea sank to the bottom and became part of the sediment.

---

*So profound is the change in the type of fossil remains found in sediments formed before and after this point in the geologic record, it was long ago selected as one of the two primary boundaries that divide the last 600 million years of Earth history into three major eras: Paleozoic, Mesozoic, and Cenozoic. It constitutes the boundary between the Mesozoic Era and the Cenozoic Era. Through radiometric dating its age has been established as 65 million years.

Why should the material in an asteroid contain so much more iridium than crustal materials on the Earth? The reason is that iridium is one of the most iron-metal-loving of all the elements. Because of this, almost all the Earth's iridium is locked up in the core. By contrast, the iridium content of a bulk asteroid would be high. It is for this reason that Luis Alvarez chose iridium.

There were a number of obvious ways to check the Alvarezes' asteroid-impact hypothesis. One was to see whether an iridium peak existed at other localities around the world where the Mesozoic–Cenozoic boundary is recorded. It did. It has now been found at 50 or so such sites. Another was to see whether other iron-metal-loving elements are enriched in the boundary sediment. They are. Platinum, gold, and silver are all enriched. Still another way was to show whether the concentration anomaly is restricted to siderophile elements. It is. More recently, quartz grains "shocked" by impact have been found in the boundary sediment. By now most scientists accept the evidence that the great extinction of life that marked the close of Mesozoic time was caused by the impact of a large extraterrestrial object. Because of this the Alvarez hypothesis gets an 8.5 on the ranking scale.

The linkage between the impact and the extinctions remains less certain. The Alvarezes postulated that the tremendous amount of dust blown up by the asteroid impact would have totally darkened the sky for several years. No plants would have been able to grow. Thus, those organisms that depended on fresh plant food would have starved and become extinct. One criticism of this explanation is that the dust would not have remained aloft long enough to have this effect. Other possibilities exist. Had the impact been in the ocean, a tremendous amount of steam would have been added to the atmosphere. The greenhouse effect of this steam could have made the Earth extremely hot. As this book was being completed, a group working under the direction of Ed Anders at the University of Chicago reported that they had found high concentrations of soot residues in the Mesozoic–Cenozoic boundary sediment. They concluded from this finding that the impact may have triggered continent-wide vegetation fires and drew an analogy to the nuclear winter scenario for the aftermath of a full intercontinental warhead exchange by the Americans and Russians.

One scientist who was not entirely convinced by the evidence amassed by the Alvarez team was Karl Turekian at Yale. So he conceived an elegant and very stringent test of their hypothesis. It involved the isotopes of the very rare iron-metal-loving element, osmium. The abundance of one of this element's seven isotopes, ^{187}Os, is being enhanced by the decay of an isotope of the element rhenium (^{187}Re with a half-life of 46×10^9 years). Working with a French scientist, Jean-Marc Luck, Turekian determined the isotopic composition of osmium associated with enriched iridium at the Mesozoic–Cenozoic boundary. For one site they found an ^{187}Os to ^{186}Os ratio of 1.65 and at another a ratio of 1.29. They compared these ratios with the average ratio for osmium from contemporary ocean sediments ($^{187}Os/^{186}Os = 7.5$) and with average values for the meteorites ($^{187}Os/^{186}Os = 1.0$). Since the value for the boundary sediment fell much closer to the meteorite value, doubting Karl had to admit that the enriched contents of the siderophile elements at this boundary could not be explained by any Earth surface source; none has so low a $^{187}Os/^{186}Os$ ratio.

The reason for the large isotopic-composition difference between whole-Earth and whole-asteroid (or comet) osmium on one hand and that for Earth-surface osmium on the other, has to do with planetary differentiation. Like potassium, the element rhenium has no comfortable mineral home in the mantle, so it moves to the crust with the first available volcanic liquid. The rhenium to osmium ratio in Earth surface rocks is about 100 times higher than that for bulk Earth and bulk asteroid material. While also prone to escape from the mantle, osmium is not so greatly enriched in the crust. Hence, the osmium in Earth surface material has received far more ^{187}Os from ^{187}Re decay than has the osmium in the Earth's interior or the osmium in asteroids or comets.

While having to admit that his findings strengthened the Alvarez case, Turekian pointed out that the ash from a megavolcanic eruption produced by material blasting its way from very deep in the Earth could have the chemical and isotopic signatures found in the Mesozoic–Cenozoic boundary sediment. While the megavolcanic hypothesis runs a weak second to the extraterrestrial-impact hypothesis, it cannot be ruled out.

Having found evidence for one such impact, two paleontologists, Raup and Sepkowski, pondered whether other such catastrophes had befallen Earth. Looking back over their paleontologic evidence, they found that episodes of high organism extinction rate seem to be spaced at intervals of roughly 26 million years. This caused a reevaluation of the assumption that asteroids were the villains. While the likelihood of impacts by sufficiently large asteroids is calculated to be several per half-billion years, there is no reason that they should occur at regular intervals. So the apparent regularity of extinction events started heads buzzing.

One idea is that the Sun has a companion star which circles it in a highly elliptical and very, very distant orbit. This as yet unseen companion was aptly named Nemesis. The idea is that Nemesis takes 26 million years to make each round about the Sun. During each round it makes a close pass to the Sun. During these "close" passes the orbits of some of the solar system's 1000 billion comets are perturbed. A few crash into the Earth. Currently an intense search of the skies goes on to see if Nemesis really exists. As of the time of publication of this book, it had not been found.

Another suggestion stems from the observation by astronomers that the Sun oscillates back and forth (like a ball on a stretched rubber band) through the disk of our galaxy. They calculate that the Sun passes through the plane of the galactic disk once each 30 million years. As the density of clouds of gas and dust is largest near the galactic plane, there is a possibility that during each pass through the disk our Sun comes close enough to one of these clouds to perturb the orbits of some of its comets. A few strike the Earth. This hypothesis received a serious blow, however, when it was pointed out that the Sun's yo-yo-like motion does not carry it beyond the heavily populated region of the disk. Hence, over the course of an oscillation only a small change occurs in the probability of a close passage to a cloud. If so, then this mechanism has no way to produce periodic impacts.

Some scientists believe that neither Nemesis nor the solar oscillation need be called upon. They challenge the evidence for a periodicity in extinction episodes. Regardless of the outcome of this debate, it seems certain that the extinction of half of the species of living organisms 65 million years ago was the result of a catastrophe of extraterrestrial origin.

Summary

While the fact that not all the solid debris in the solar nebula was collected by the nine planets offers us more information regarding our heritage, it also poses a threat. Unlike the information-laden meteorites that fall harmlessly to Earth, on rare occasions the Earth appears to have been struck by kilometer-sized asteroids or comets which dealt devastating blows. These disasters surely punctuated the evolution of life. It is possible that on some habitable planets the development of life was snuffed out by such an impact.

Supplementary Readings

Origin of the Earth and Moon, *by A.E. Ringwood, 1979, Springer-Verlag.*

A comprehensive discussion of the origin of the Moon written for the specialist.

Toward a theory of impact crises, *by Walter Alvarez, 1986, EOS, American Geophysical Union.*

A summary of the evidence in support of the impact hypothesis for the mass extinction at the Mesozoic-Cenozoic boundary.

The Barnard Glacier

Chapter Seven

Making It Comfortable:

Running Water and Temperature Control

The temperature of a planet's surface depends on the luminosity of the star it orbits and on the planet's distance from this star. It also depends on the reflectivity of the planet's surface and on the "greenhouse power" of its atmosphere. Earth and Venus exemplify the importance of the latter of these influences. Although nearly equal in size and bulk composition, the Venusian ground surface is 400 degrees centigrade warmer than Earth's. The reason for this difference is that on Venus most of the carbon is in the atmosphere as CO_2. The thermal blanket provided by this CO_2 raised the Venusian surface temperature by some 400 degrees centigrade. By contrast, on Earth most of the carbon is stored in sediments as carbonate minerals and organic residues. The amount of CO_2 in the atmosphere of Earth is 350,000 times less than that in the atmosphere of Venus.

The CO_2 content of the Earth's atmosphere has probably changed with time in response to changes in the rates and patterns of the convective motions in the Earth's mantle. These motions cause sections of the Earth's sedimentary cover to be dragged into the mantle where they are heated to the point where sediment-bound carbon is converted to CO_2 gas, and this CO_2 is carried back to the surface with volcanic fluids. Changes in the rate and pattern of the mantle's flow undoubtedly have led to changes in the rate at which CO_2 is added to the atmosphere. Hence, the Earth's climate is linked to its tectonics.

The climate on Earth is also sensitive to details of planetary architecture. Because of its equatorial bulge and its tilted rotation axis, the Earth undergoes a top-like precession once every 20,000 years. Because of the gravitational tug of the major planets, the Earth's orbital tilt and orbital shape change cyclically on the time scales of 40,000 and 100,000 years. Scientists now believe that the changes in the seasonal and latitudinal distribution of the sunlight reaching the Earth which accompany the orbital cycles cause the Earth's polar ice caps to wax and wane.

Introduction

We have yet to cover the most important attributes of a planet bearing on its habitability. The matters we have dealt with so far only set the stage for the discussion of those characteristics most important to living organisms. What fixes the planet's water supply? What sets the temperature of its surface? What supplies oxygen gas? What determines the planet's energy reserves? What gives the planet beauty? In a nutshell, what makes it habitable?

There are, of course, no simple answers to these questions. The habitability of a planet is in part determined by its nebular heritage which sets its size, orbit, spin, and chemical composition. It is in part determined by the evolution of its interior and crust. As we shall see in this chapter, it also critically depends on what happened to its volatiles after they reached the planet's surface. What fraction escaped to space? What fraction took up residence in the atmosphere? What fraction in the ocean? What fraction in sediments? What fraction in ice caps? In this connection we shall see that life has modified our planet's environment. It reshuffled the volatiles between atmosphere, ocean, and sediment. It adds O_2 to the air. It creates coal, oil, and natural gas.

To unravel the roles of nebular heritage, planetary evolution, volatile residency, and life is a tall order. We cannot hope to do it justice here. Even if space were available, many of the answers are not. Instead, we will focus on a few major issues.

Water

For life of any complexity to develop on a planet, liquid water must be available in abundance. The most stark differences we see from region to region on Earth's continents relate to water. In regions of the continents where there is much rain we have lush forests teeming with all manner of living organisms. Where there is little rain we have deserts sparse in life. Where only snow falls we have ice caps nearly free of life. These stark contrasts are found on a planet whose surface is 70 percent covered by liquid water!

Four requirements for a suitable water supply are as follows: 1) the planet must have captured enough water to make a sizeable ocean; 2) this water must have migrated from the planet's interior to its surface; 3) it must not have been lost to space; 4) it must be largely in liquid form (the mean surface temperature must lie above the freezing point of water and below the boiling point of water).

The Earth is about one-half of 1 percent by weight water. To become so required that it capture only one atom out of every 3 million hydrogen atoms available in the pool of matter from which it formed. The factors that led to this particular capture efficiency are not understood. Based on the observation that the Earth is depleted in potassium and other moderately volatile elements, our planet would not be expected to have any water. To explain its presence requires that the chunks of matter that went into its construction were a mixed bag. The majority of these pieces must have been more depleted in volatiles than are ordinary chondrites. A minority must have resembled carbonaceous chondrites in that they carried sizable components of minerals containing bound water. However, as we have little information to go by, there is no purpose in pondering this matter. Somehow Earth got just about an ideal amount of water to support life. The situation would have been very different had the Earth captured ten times more or ten time less water. With ten times more water there would be little land area; the ocean would likely cover 99 or more percent of the planet's surface. With ten times less water the oceans would be tiny and the deserts vast!

As we have learned from the ^{129}Xe story (see Chapter Five), the Earth's xenon must have been released from the body of the planet very early in the Earth's history. At the high temperatures that prevailed during this core formation, water would have been in gaseous form. Thus, it should have migrated to the surface along with xenon.

Once at the surface, any gaseous substance has the opportunity to evaporate into space. The likelihood that a molecule will escape depends both on the gravity of the planet (that is, its mass and radius) and on the mass of the molecule itself. These dependences are extremely strong. A factor of two differences in the mass of the gas molecule corresponds to a change of several orders of magnitude in the likelihood of escape to space. For example, the mean time required for a helium atom (4 nuclear particles) to escape from the Earth is about 1 million years, while that for neon (20 nuclear particles) is many billions of years. Hence, the Earth's atmosphere loses helium to space just as fast as helium is gained through outgassing of the interior. By contrast, the Earth has held onto nearly all its neon.

The same physical laws that allow light atoms to escape preferentially also apply to light planets. The ability of a planet to hold its gases is strongly dependent upon its gravity. While Earth and Venus are massive enough to hold all but the lightest gases, the Moon and Mercury have insufficient gravity to hold any gas. Thus, Earth and Venus have substantial atmospheres while the Moon and Mercury have none at all.

This very strong dependence of escape probability on molecular mass is a natural consequence of gas kinetics. For a molecule to escape the Earth's gravity it must, like NASA's rockets, have an outward velocity of 25,000 miles per hour (11 km/sec). This is far, far greater than the average velocity of molecules residing near the "top" of our atmosphere. A molecule can only achieve the needed velocity by the remotest chance. The probability that a heavy molecule by chance achieves the required velocity is many powers of 10 lower than is the case for a light molecule.

The ambient velocity of molecules near the "top" of a planet's atmosphere is a measure of the temperature of this part of the atmosphere. This temperature is generally considerably higher than that at the planet's surface. For the Earth the temperature of the top of the atmosphere is thought to be about 1500° Kelvin, while that at the surface is only about 300° Kelvin. This higher temperature greatly increases the probability for molecular escape. Were the Earth's atmosphere to have a temperature of 300° Kelvin from bottom to top not even helium would escape at a significant rate.

We have a means to estimate the current escape probability of helium from Earth. This method involves a comparison between the total number of helium atoms in the atmosphere and the number of helium atoms that leak into the atmosphere from the Earth's interior each year. The number in the air is, of course, obtained from the mass of the atmosphere and its helium content. Helium is generated within the Earth by the decay of ^{238}U, ^{235}U, and ^{232}Th. By measuring the rate at which the temperature increases with depth in mine shafts, bore holes, and the like, geophysicists have been able to calculate how much heat escapes from the Earth's interior each year. This heat is produced by the radiodecay of the same three isotopes that produce helium and by the decay of ^{40}K. As the ratio of K to U for the Earth as a whole is known (see Chapter Three), the ratio of heat production to helium production in the Earth's interior can be calculated. Knowing how much heat is getting out each year, we can calculate how much helium is getting out each year. It turns out that helium is escaping from the Earth's interior at more or less the same rate it is being created by the decay of the uranium and thorium (see Figure 7-1). The number of atoms escaping in a year is about one-millionth the number of helium atoms currently in the atmosphere. This suggests that helium atoms require on the average 1 million years to accomplish their escape from the atmosphere to space. Having established the helium escape time, the escape time for the other gases can be calculated from molecular theory.

As they are are only half as massive as helium, hydrogen molecules have an escape time considerably less than 1 million years. Fortunately, H_2 is a very rare gas in our atmosphere (see Table 7-1). Those H_2 molecules generated by bacteria living in soils survive only a few years in the atmosphere before being converted to water ($2H_2 + O_2 \rightarrow 2H_2O$). Because of this low abundance, the loss of H_2 gas has not taken a serious toll on the Earth's hydrogen (and hence, its water) reserves.

There is, however, another route by which a planet loses water. High in the atmosphere ultraviolet light breaks H_2O molecules apart, creating free H atoms. These free hydrogen atoms are very vulnerable to escape. The reason why this process has not decimated the Earth's water reserves is that only

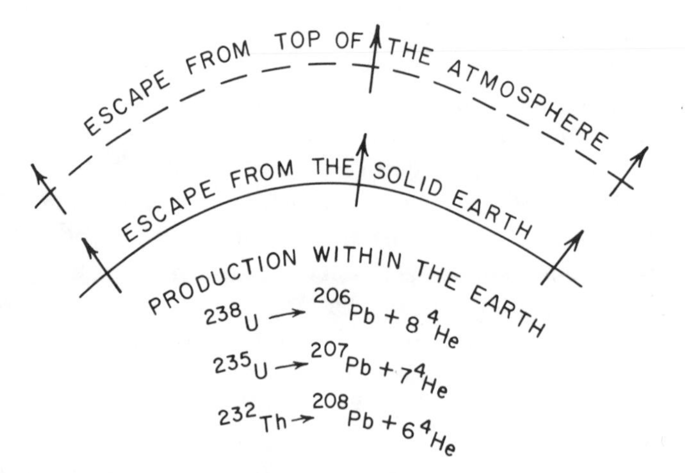

Figure 7-1. Earth history of helium atoms: ^{4}He atoms are generated within the Earth's crust and mantle by the decay of uranium and thorium. By measuring the amount of heat escaping from the Earth, we have a fairly good idea of how much uranium and thorium are present in the Earth. Thus we know the rate at which ^{4}He generation occurs. After a billion or so years of entrapment in the solid Earth, the average helium atom manages to reach the surface, where it resides an average of 1 million years before "evaporating" from the top of the atmosphere. All the helium atoms manufactured by radiodecay within the Earth are eventually lost to space.

Table 7-1. Composition of Earth's Atmosphere:*

GAS NAME	GAS FORMULA	PERCENT BY VOLUME
NITROGEN	N_2	78.08
OXYGEN	O_2	20.95
ARGON	Ar	0.93
CARBON DIOXIDE†	CO_2	0.034
NEON	Ne	0.0018
HELIUM	He	0.00052
KRYPTON	Kr	0.00011
XENON	Xe	0.00009
HYDROGEN	H_2	0.00005
METHANE†	CH_4	0.0002
NITROUS OXIDE†	N_2O	0.00005

*In addition, the atmosphere contains water vapor (H_2O) in variable amounts (up to 2 percent for warm air and down to a few parts per million for very cold stratospheric air). Water is also a greenhouse gas.

†Greenhouse gases.

a very tiny fraction of the Earth's water resides in the stratosphere. As tallied in Figure 7-2, the bulk of the water is in the ocean, sediments, and ice. At any given time only about one H_2O molecule in 10^5 is in the atmosphere. Furthermore, almost all the atmosphere's water resides in the troposphere. The very low temperature (-90° C) at the base of the stratosphere thoroughly dries any air that is transferred from the troposphere to the stratosphere (the water vapor crystallizes to snow which falls back into the troposphere). Because of this the Earth's hydrogen escape hatch is quite small. Over the last 4.6 billion years only a very small fraction of our water has been lost!

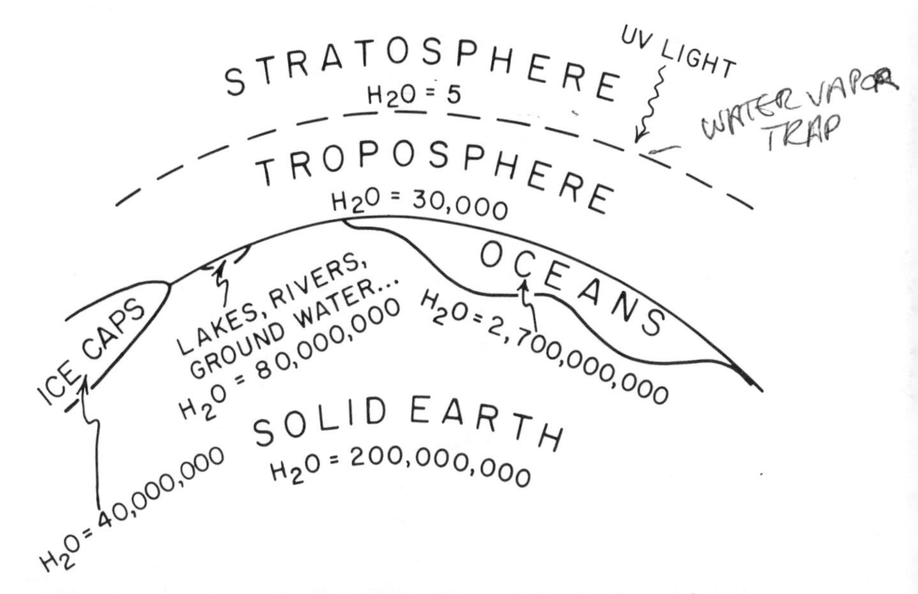

Figure 7-2. Most of the Earth's hydrogen is in the form of water: Roughly nine-tenths of this water resides in the ocean. The majority of the remaining one-tenth is trapped in solids making up the mantle and crust. The Earth's fresh waters (lakes, rivers, ground waters, etc.) make up only about 3 percent of the total. Its ice caps make up only 1.5 percent of the total. Only a very small fraction of our water is at any given time stored in the atmosphere as vapor. This vapor is confined almost entirely to the lower, well-stirred part of the atmosphere (referred to by meteorologists as the troposphere). Only two of every billion water molecules reside in the stratosphere. This is important, because only that water present in the stratosphere has a chance to be dissociated by ultraviolet light. The numbers represent the amounts of water in each reservoir (in units of 10^8 grams).

Surface Temperature

Beyond having started with an appropriate amount of water, having outgassed this water to its surface, and having been able to hold onto this water, the remaining requirement is for the Earth to have achieved a temperature allowing this water to be largely in liquid form. Our water would be of little use to life were it tied up in massive ice sheets. Nor would it be of use were it all in the atmosphere as steam.

The temperature of a planet's surface is not only dependent on the amount of light it receives from its star; it also depends on the reflectivity of its surface and on the content of so-called greenhouse gases in its atmosphere. Only if a planet had the surface properties of what physicists refer to as a black body would its temperatures be fixed solely by the amount of light reaching its surfaces. To qualify as a black body the surface of the planet must be nonreflecting; that is, all the sunlight reaching it must be absorbed and later reradiated as planet light (see Figure 7-3). Also, there can be no gases in its atmosphere that absorbed outgoing planet light. Were the Earth a black body it would have a mean surface temperature of about 5° C (see Table 7-2).

The Earth is not a black body. First of all, its clouds, its ice caps, and its deserts reflect back to space a sizable portion (≃ 33 percent) of the sunlight impinging upon the Earth. The sunlight reflected back into space plays no role in maintaining the temperature of the Earth's surface. Were this the only deviation from the black-body situation, the Earth's surface temperature would average −20° C (all water would be frozen). However, the Earth's atmosphere contains molecules made up of three or more atoms. Molecules of this type are able to capture packets of outgoing infrared light (see Figure 7-4). Important in the Earth's atmosphere are water vapor (H_2O), carbon dioxide (CO_2), methane (CH_4), and nitrous oxide (N_2O). The capture of outgoing earth light by these gases warms the planet. For Earth the greenhouse warming more than compensates for the sunlight lost through reflection. The Earth's mean surface temperature is 10° C warmer than if it were a black body (see Table 7-2).

Thus, to assess a planet's temperature, we need to know not

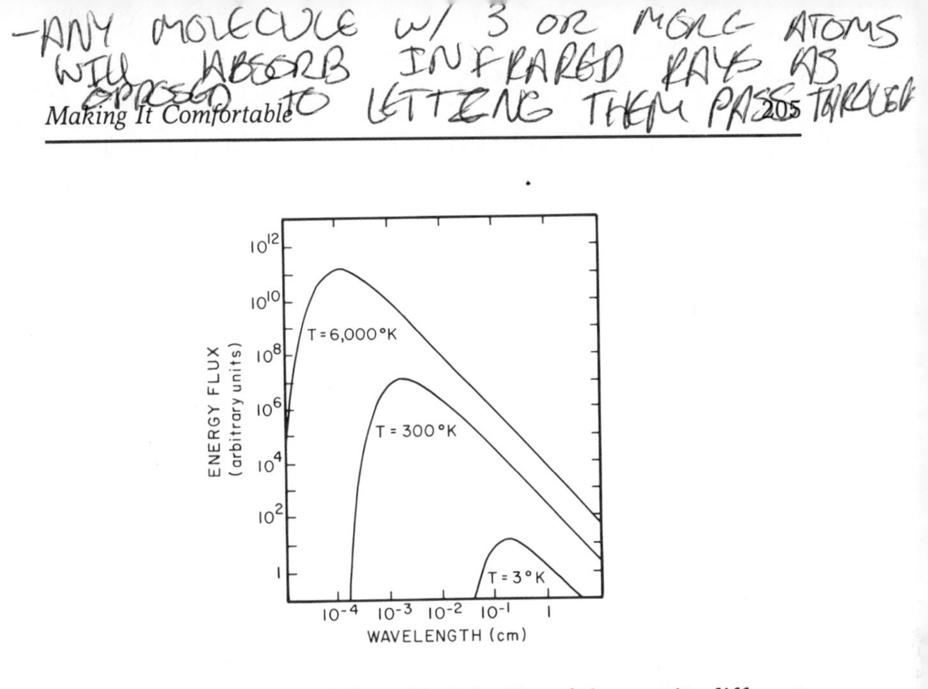

Figure 7-3. Light emitted from black bodies of three quite different temperatures: The hotter the body, the more energy it emits (the 6000° K body gives off about 10,000 times more energy per unit area than the 300° K body and 100 million times more per unit area than the 3° K body). The wavelength representing the peak of energy emission for a star whose surface temperature is 6000° K lies in the visible range, that for a planet whose surface temperature is 300° K lies in the infrared range, and that for a universe whose background glow temperature is 3° K lies in the microwave range.

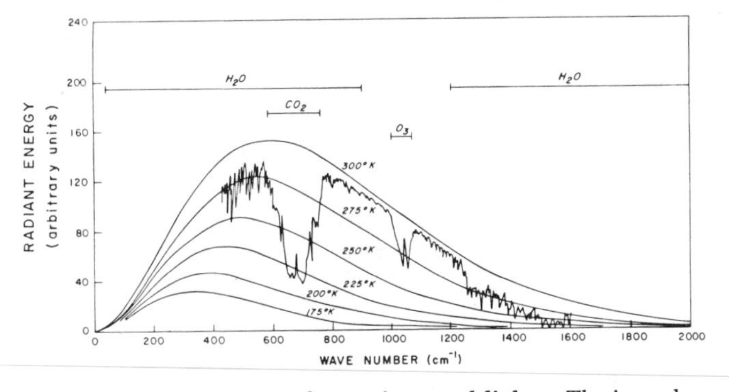

Figure 7-4. Absorption of outgoing earthlight: The jagged curve shows an actual spectrum of earthlight leaving the top of the atmosphere over the island of Guam. For comparison, the smooth curves show the expected spectra if there were no greenhouse gases in the atmosphere; these curves are drawn for a series of temperatures. The wiggles and large dips are the result of absorption of the light by water, carbon dioxide, and ozone in the atmosphere. As can be seen, the dip created by CO_2 is especially prominent. The wave number is a measure of the frequency of the radiation. The higher the wave number, the higher the frequency.

Table 7-2. Summary of the factors influencing the surface temperatures of the terrestrial planets:

	Mass of atmosphere $\frac{kg}{cm^2}$	Distance from Sun 10^6 km	Solar energy received $\frac{10^6 \text{ ergs}}{cm^2 \text{ sec}}$	Black-body temperature C°	Fraction sunlight reflected	Reflective cooling C°	Greenhouse warming C°	Actual surface temperature C°
Mercury	0	58	9.2	+175	.06	-5	0	-
Venus	115*	108	2.6	+55	.71	-84	+460	+430
Earth	1.03 †	150	1.4	5	.33	-25	+35	+15
Mars	0.016*	228	0.6	-50	.17	-10	+15	-45

*Mostly CO_2
†Mostly $N_2 + O_2$

only how much sunlight it receives but also the reflective properties of its surface and the amounts of infrared-light-absorbing gases in its atmosphere.

The reflectivity of a planet has much to do with the amount and state of its water. Ocean water is not very reflective; ice is highly reflective; clouds are moderately reflective. Plants absorb nearly all the light they receive; there is no glare in a forest. By contrast, about half the light reaching vegetation-free portions of the continents is reflected. As we have already pointed out, the extent of plant cover depends on the amount of rainfall.

The greenhouse capacity of a planet also has much to do with the state of its water. For today's Earth, water vapor is the most important absorber of outgoing infrared light. The warming produced by this greenhouse capacity by chance just about compensates for the cooling produced by the reflectivity of the Earth's clouds.

It need not be this way. Imagine that by some magic the Sun's energy output could be turned down for a long enough time for the continents to become snow-covered and the ocean to freeze. Then the Sun is turned back up again to its present output. The ocean would remain frozen, because snow is so reflective that most of the energy reaching us from the Sun would "bounce" off. The Earth would remain cold!

Or imagine that by some magic the Sun's energy output was turned way up for long enough to permit the oceans to boil away, producing a dense atmosphere of steam. Then the Sun is turned back down again. It would remain very hot on Earth. The immense greenhouse effect of the steam would maintain the planet's ground temperature so high that the steam could not recondense. The Earth would remain hot!

So we see that the situation is not so simple. While the Earth's black-body temperature is such that its water should be mainly in liquid form, given the high reflectivity of ice, it could just as well be frozen over; or, given the high greenhouse capacity of steam, it could be scalding.

A Lesson from Venus

Carbon dioxide also plays an extremely important role in establishing a planet's surface temperature. In the Earth's atmosphere CO_2 is second only to water in its greenhouse capacity. What controls its amount? In the nebula most of the carbon was in the form of methane gas. As such it was lost. The Earth, however, somehow managed to capture about one in every 3000 carbon atoms available in the nebular material from which it formed. Most of this carbon is now at the Earth's surface. Only a tiny fraction, however (about 60 atoms out of every million), is currently in the atmosphere as CO_2. The bulk resides in sediments, part as calcium carbonate (called limestone by geologists), and part as organic residues (called kerogen by geologists). Were all the Earth's carbon in the form of CO_2, its amount would exceed that of the current N_2 and O_2 by a factor of about 100. The pressure exerted by this CO_2 atmosphere would be a staggering 100 times greater than now (similar to the pressure experienced by the hull of a nuclear submarine submerged to a depth of 1 kilometer).

We have a dramatic reminder that the situation with regard to CO_2 could well be different. The reminder is the planet Venus, which has a whopping big atmosphere made almost entirely of CO_2. The greenhouse effect of this CO_2 atmosphere gives Venus the scalding surface temperature of 430° C. As Venus and Earth have nearly the same size, nearly the same density, and nearly the same K/U ratio, it seems reasonable that they also started with the same volatile content. Indeed, the fact that the amount of carbon in the CO_2 of the Venusian atmosphere is about the same as the amount of carbon locked up in limestone and kerogen on the Earth's surface provides evidence that this is true.* Thus, Venus has the conditions that would prevail on Earth

* Because Venus is so hot, it surely has no life and hence no kerogen. Also, $CaCO_3$ would decompose under these conditions, releasing its carbon as CO_2 gas. Hence, it is likely that nearly all the carbon on the Venusian surface resides as CO_2 gas in its atmosphere.

if all the CO_2 locked up in limestone and kerogen were to be released as CO_2 to the atmosphere.

However, when comparing Earth and Venus a problem arises in connection with water. If Venus started with the same component of volatiles as Earth it should have a sizable ocean (or rather, at its high temperature, an atmosphere dominated by steam).* Not only is the atmosphere of Venus not dominated by steam, but water vapor is barely detectable.

Most scientists believe that the hydrogen initially present on Venus as water escaped to space. In the very hot Venusian atmosphere, water vapor would be effectively transported to the "top." Here it could be disassociated by ultraviolet light to form hydrogen atoms which would then escape. The "left behind" oxygen atoms would be stirred back down through the atmosphere to the surface of the planet where they would gradually convert the FeO in the hot Venusian crust to Fe_2O_3.

Evidence in support of this hypothesis was obtained when an unmanned American space probe was dropped into the atmosphere of Venus. Before this probe was rendered inoperative by the high temperatures of Venus, it measured and radioed back to Earth the isotopic composition of the trace amount of water present in the Venusian atmosphere. The astounding finding was that the amount of deuterium (^2H) in Venusian hydrogen is 100 times higher than the amount in Earth hydrogen. Because of their two-fold larger mass, ^2H atoms have a much lower escape probability than do ^1H atoms. Hence, the escape of hydrogen from Venus would tend to enrich the residual water in deuterium. While the observed hundred-fold enrichment of deuterium does not prove that Venus once had as much water as Earth, it can only be explained if Venus once had at least a thousand times more water than it does now!

Thus, it is entirely possible that Venus and Earth started with roughly the same volatile ingredients. Earth for some reason

---

* If the Earth were heated to the point where its ocean was converted entirely into steam, this steam would exert a pressure about 270 times that of the present Earth atmosphere.

evolved in such a way that it kept its carbon safely buried in sediments. In so doing it avoided the disastrous consequences of the so-called runaway greenhouse effect. Venus, on the other hand, slipped at some point and let CO_2 build up in its atmosphere. This buildup led to high temperatures which terminated life (if indeed life ever had a foothold on Venus). It is hard to imagine how a planet once in this very hot state could ever become cool again.

Not only did the Earth somehow manage to avoid the runaway greenhouse trap, but it also avoided the trap of becoming "white." Had either a runaway greenhouse or a runaway reflectivity disaster befallen the Earth at some time in the past, we would see clear evidence of this event in the sedimentary record. At least over the last 3 billion years, during which the record is reasonably complete, nothing of the sort is found. Sediments deposited by water are found in all rock sequences. Hence, the Earth's water was neither all ice nor all steam. Planet Earth has steered a very careful course as far as climate goes. Its mean surface temperature has remained above the freezing point and below the boiling point of water.

We have no way to learn about the history of Venus. It is difficult to imagine that astronauts will ever prance about its surface as they did on the Moon. While the Russians have managed to land several unmanned space probes on the hot surface of Venus, these vehicles survived the hostile conditions only long enough to radio back information about the temperature, pressure, and composition of the Venusian atmosphere and (as we learned in Chapter Three) about the potassium to uranium ratio in the rock surface upon which the probe landed. Radar beams bounced off Venus tell us that its surface has large topographic features which smack of active tectonics. Unfortunately, these findings provide no clues as to why the climate of Venus went haywire. Was it because of the extra sunlight that Venus receives? Was it because life never got started on Venus? Was it because the initial component of water on Venus was much smaller than that on Earth? Or did God put Venus there as a symbol of what can go wrong if a planet is mismanaged?

The Earth's Thermostat

While the Earth's temperature appears to have remained within the range 0° to 100° C for much of geologic time, it has not been constant. During the last 10 or so million years it has gone through a cold spell. Today continental ice sheets cap Antarctica and Greenland. This ice cover was substantially larger during the eight or so major glaciations of the last million years. By contrast, during Cretaceous time, 100 million to 65 million years ago, the Earth was warm enough that there were no standing ice caps. Temperate climates extended all the way to the poles! Evidence for glaciation reappears in the time interval 240 million to 400 million years ago and again in the time interval 2100 million to 2500 million years ago. What factors are responsible for these swings toward cold? Could there exist a type of thermostat which has prevented these excursions from engulfing the entire Earth?

A very interesting argument can be constructed to show why the Earth never became frozen. To understand it we will have to delve a bit into the chemical cycles that occur in the Earth's crust, soils, and ocean. To keep these discussions as simple as possible, we will restrict our thinking to a hypothetical planet designed to have the essential features of Earth but not the host of complicating details (see Figure 7-5).

Our hypothetical planet will have a solid crust consisting of wollastonite, which has the chemical formula $CaSiO_3$. Chemical activity in the soils capping this crust leads to the following reaction:

$$3H_2O + 2CO_2 + CaSiO_3 \rightarrow Ca^{++} + 2HCO_3^- + H_4SiO_4^\circ$$

Carbon dioxide gas dissolved in soil waters causes wollastonite to dissolve, yielding ions of calcium, bicarbonate, and silicate. These ions percolate with the soil water through the underlying rock to a nearby stream and eventually to the sea. In the sea, organisms use these ions to make their shells. Some organisms make their shells of calcite. They carry out the reaction

$$Ca^{++} + 2HCO_3^- \rightarrow CaCO_3 + H_2O + CO_2$$

Others form opal. They carry out the reaction

$$H_4SiO_4^\circ \rightarrow SiO_2 + 2H_2O$$

The calcite ($CaCO_3$) and opal (SiO_2) hard parts fall to the sea floor (along with detrital grains of wollastonite eroded from the continents) to form sediment.

This sediment accumulates on the great crustal plates until they reach a zone of convergence. Here some of the sediment is subducted into the mantle. At the high temperature of the Earth's interior the calcite reacts with the opal to yield wollastonite and carbon dioxide gas,

$$SiO_2 + CaCO_3 \rightarrow CaSiO_3 + CO_2$$

The wollastonite becomes part of the crust, completing its cycle. The CO_2 escapes to the atmosphere, completing its cycle.

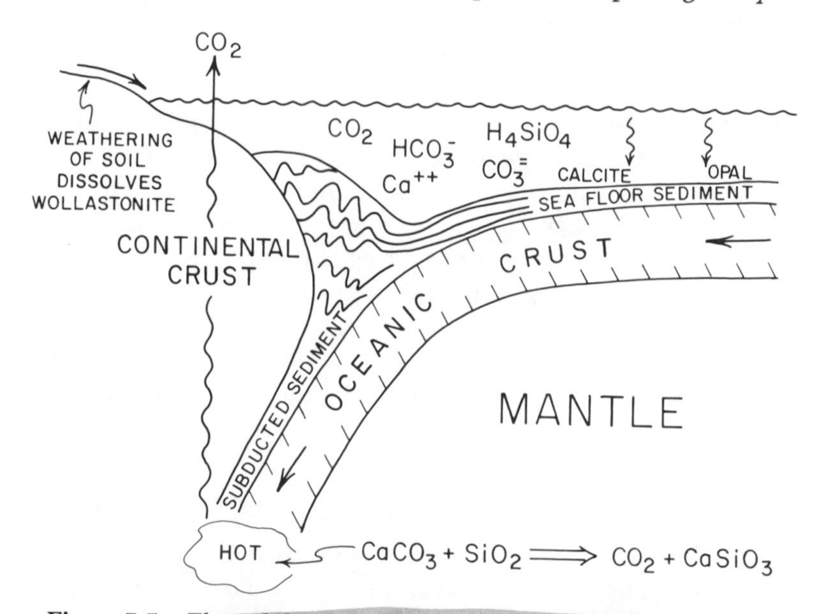

Figure 7-5. The subduction of oceanic crust beneath the continents carries part of the blanket of sediment deep into the Earth's mantle: Here they are heated and metamorphosed. During this process any carbonate minerals contained in the sediment are broken down, releasing CO_2. This CO_2 migrates back to the Earth's surface and rejoins the ocean-atmosphere reservoir. Eventually it is recombined by marine organisms with calcium in the mineral calcite. This calcite is buried on the sea floor and starts another trip toward a subduction zone.

So much for the background. The interesting aspect of this cycle is how it influences the CO_2 content of the atmosphere. The basic driving mechanism for the cycle is the plate motions that carry sediment from the Earth's surface to regions in its interior hot enough to cause opal and calcite to react with one another, releasing CO_2 gas. This sets the rate at which CO_2 is added to the atmosphere–ocean system. If at any given time CO_2 was not removed from the ocean (through calcite burial in the sediment) as fast as it was being added to the atmosphere–ocean system, then the CO_2 content of the atmosphere would steadily increase. On the other hand, if organisms were to remove calcite from the ocean too fast, the CO_2 content of the ocean–atmosphere system would steadily decrease.

Somehow a balance between the supply of CO_2 to the ocean–atmosphere reservoir and removal of CO_2 from this reservoir must be achieved. The key factor in this balance is that in order to form calcite, organisms need calcium as well as CO_2. Thus, calcite can accumulate in marine sediments no faster than calcium is being made available through the chemical reactions taking place in continental soils. The rate of these chemical reactions depends on the temperature of the soil (all chemical reactions go faster when the reactants are heated), on the rainfall (the more water that runs through the soils, the more reaction that occurs), and on the CO_2 content of soil waters (since CO_2 is an acid, the more CO_2 in the water, the faster the soil minerals dissolve).

Now comes the clever part of the argument. As stated above, if CO_2 was added to the atmosphere faster than it was removed by calcite deposited in deep sea sediments, then the CO_2 content of the atmosphere would increase. This would:

1) make the planet warmer (because of the increased greenhouse effect);
2) make the planet wetter (warmer air holds more water vapor and hence makes more rain); and
3) make the CO_2 content of soils higher.*

* This happens for two reasons. First, the more CO_2 in the air, the more will dissolve

Thus, a rise in atmospheric CO_2 content would increase the rate at which calcium dissolves from the continents and thereby permit calcite to accumulate more rapidly in marine sediment. Eventually the calcite production rate would become great enough so that CO_2 could be removed from the atmosphere–ocean system as fast as it was added. The CO_2 buildup would be stemmed.

With this background in mind it is easy to envision what would happen if the Earth's waters were to freeze. There would be no marine organisms to make calcite. Nor would there be any chemical erosion. Thus the CO_2 released from the Earth's hot interior would build up in the atmosphere until the temperatures became warm enough to melt the ice. Only then could erosion and calcite formation begin once again.

Does this mean that no planet in the universe endowed with carbon could ever freeze? No. We have already pointed out that if a planet is too small its CO_2 would escape to space. Also, if a planet were too far from its sun it might be cold enough that CO_2 itself would freeze. There is a suspicion that CO_2 caps form on Mars. So while this CO_2 thermostat is effective on Earth, it is by no means universal.

There appears to be no equivalent thermostat preventing Earth from getting too warm. The situation on Venus speaks against such a control system.

Ice Ages

While managing to avoid the reflectivity and greenhouse disasters, the Earth's temperature has not remained constant. The record preserved in sediments deposited on the sea floor tells us that important changes have occurred even on short time scales. The most dramatic messages are those for the last million years. Over this time interval there has been one major glacial

---

in soil waters. Second, the more rain, the more plant cover; the more plant cover, the more root respiration; the more root respiration, the more CO_2 in the soil. The second is probably considerably more important than the first.

cycle every 100,000 years. Although not entirely regular, these cycles can be characterized by gradual buildups of ice caps on the North American and European continents. Each of these buildups was suddenly terminated by warmings which brought climate rapidly back to full interglacial conditions. The contrast between the extent of ice cover during peak glacial and peak interglacial times is shown in Figure 7-6.

While the geographical extent of these vast continental glaciers is determined by the deposits they left behind, the best record of their history comes from sediments on the sea floor. This record is preserved in the shells of little snail-like organisms, called benthonic foraminifera, which inhabit the sea floor. The ratio of ^{18}O atoms to ^{16}O atoms in the $CaCO_3$ of their shells records the ratio of ^{18}O to ^{16}O atoms in sea water. The important point is that the ratio of ^{18}O to ^{16}O atoms in sea water changes with the growth and retreat of the ice caps. Because of separations between water molecules having ^{16}O atoms and water molecules having ^{18}O atoms during condensation of water in clouds, the snow falling on ice caps has a $^{18}O/^{16}O$ ratio several percent lower than sea water's. The ^{18}O deficiency in the ice accumulated in the continental glaciers must be mirrored by an ^{18}O excess in the water remaining behind in the sea (see Figure 7-7). Because of this, during times of glaciation the $^{18}O/^{16}O$ ratio in sea water was somewhat higher than it is today. Shown in Figure 7-8 is the $^{18}O/^{16}O$ record obtained from shells picked from various depths in sediment cores taken from the sea floor. As amazing as it may seem, these tiny organisms provide us with a record of continental ice volume. By using radioisotope hourglasses akin to those used to date meteorites, a time scale has been obtained for these records.

The realization by geologists that the Earth's climate has undergone dramatic cycles brought with it a curiosity as to what caused these shifts. From the very first, the prime candidate has been cyclic changes in the Earth's orbit about the sun. Although when averaged over very long periods of time the Earth's path around the sun does not change, its orbit does show some repetitive deviations from this mean shape.

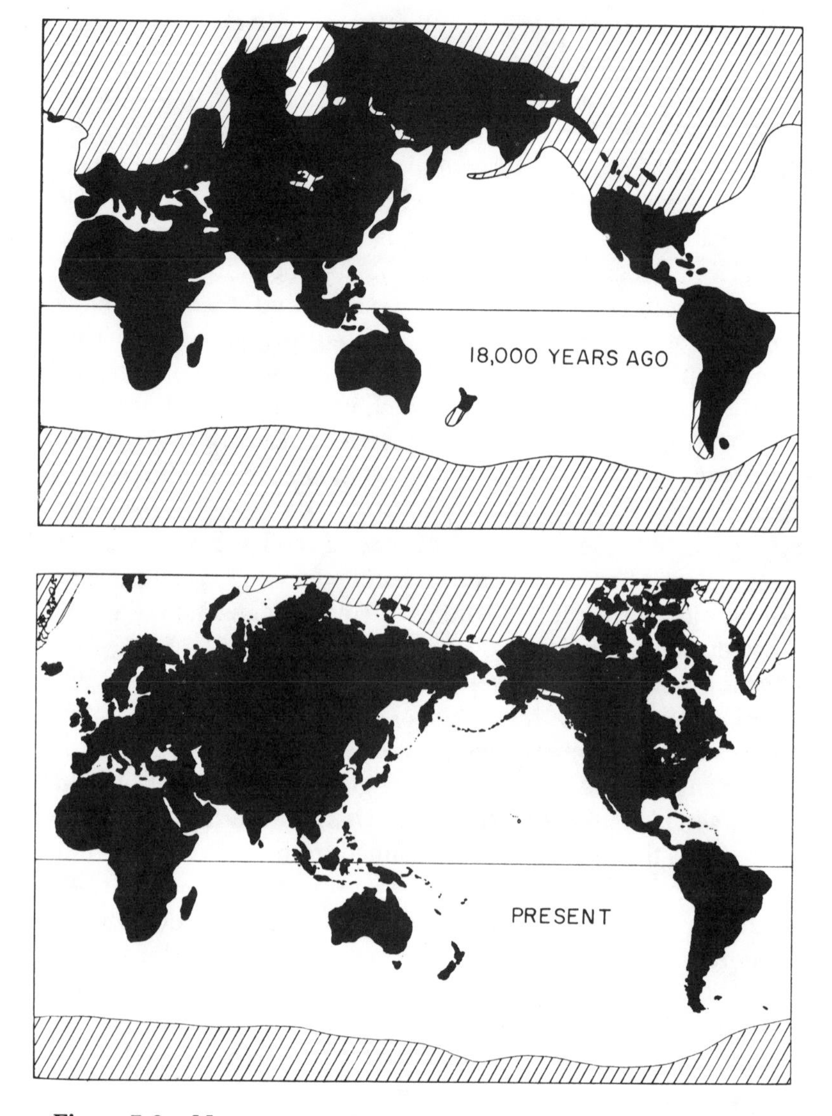

Figure 7-6. Maps contrasting the extent of ice cover during the peak of the last period of glaciation 18,000 years ago with that which exists at present: Part of this ice is continental and part floats in the sea.

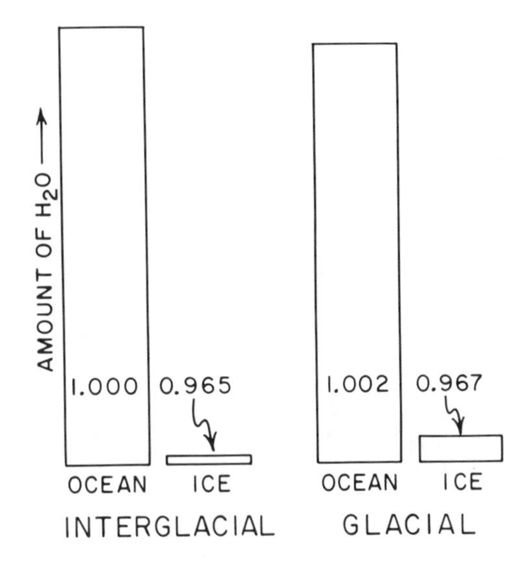

Figure 7-7. Comparison of the amounts of H_2O in the ocean (as water) and in the continental glaciers (as ice) during times of full interglaciation (like today) and during times of full glaciation (like 18,000 years ago): The heights of the bars indicate the amount of water in each reservoir. The numbers associated with the bars are the $^{18}O/^{16}O$ ratio for the H_2O divided by the $^{18}O/^{16}O$ ratio in today's ocean. Since the ice is about 3.5 percent depleted in ^{18}O compared to sea water, the growth of glaciers causes the ocean to become slightly enriched in ^{18}O.

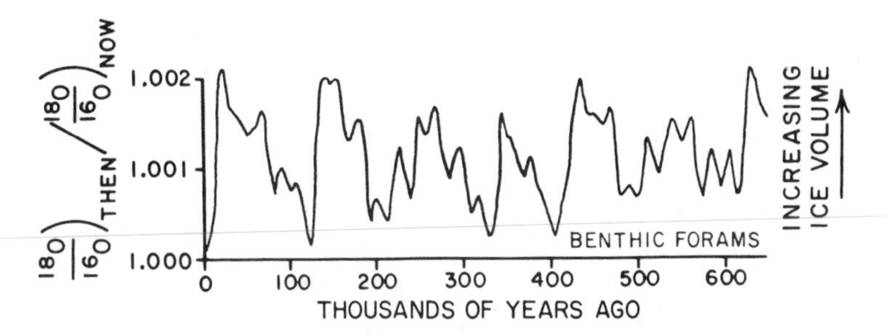

Figure 7-8. The $^{18}O/^{16}O$ record for benthic foraminifera from deep-sea sediments: Higher-than-present $^{18}O/^{16}O$ ratios correspond to times when more water was locked up in continental glaciers. The time scale was obtained from radioisotope measurements. Although not a regular progression, the times of rapid deglaciation follow one another at roughly 100,000-year intervals.

The importance of these orbital changes to climate lies in the fact that they alter the contrast between the seasons. Changes in the distribution between summer and winter of the sunlight received at any given location on the Earth's surface are caused by two characteristics of its orbit. The first is associated with the tilt of the Earth's spin axis with respect to the plane of its orbit about the Sun (see Figure 7-9). The Earth's spin axis does not stand straight up. Because of this tilt, at one point in its orbit, the northern hemisphere faces the Sun. Six months later the southern hemisphere faces the Sun. If the Earth stood straight up its equator would always face the Sun and there would be no tilt-induced seasons. The larger the tilt, the greater the seasonal difference in the amount of radiation received.

Also shown in Figure 7-9 is the second source of seasonality. The Earth's orbit is slightly out of round. In geometric terms, the Earth's orbit is an ellipse rather than a circle. As you remember, circles can be drawn by tying a string to a piece of chalk. One end of the string is held at the center (focus) of the circle; to draw the circle the chalk is swung around the focus in an arc. An ellipse has two foci. In order to draw one, both ends of the string are pinned down, one end at each focus. The chalk is not attached; rather it is nested freely in the vee made by stretching the string. The arc drawn by swinging this vee around is an ellipse. The farther apart the foci, the more eccentric the ellipse (that is, the more it deviates from circularity). No planet or moon has a perfectly circular orbit. All are elliptical (see Table 3-2, Chapter Three). The laws of gravity require that one of a planet's two foci correspond in location to the Sun.

The consequence of the elliptical shape of the Earth's orbit is clear. The Earth–Sun distance changes over the course of the year. When closer than average to the Sun, the Earth receives more radiation; when farther than average, it receives less radiation.

Shown in Figure 7-10 is the relationship between the influences of the two seasonal cycles. Currently in the northern hemisphere they oppose one another. The reason is that the Earth points its northern hemisphere toward the Sun as it goes

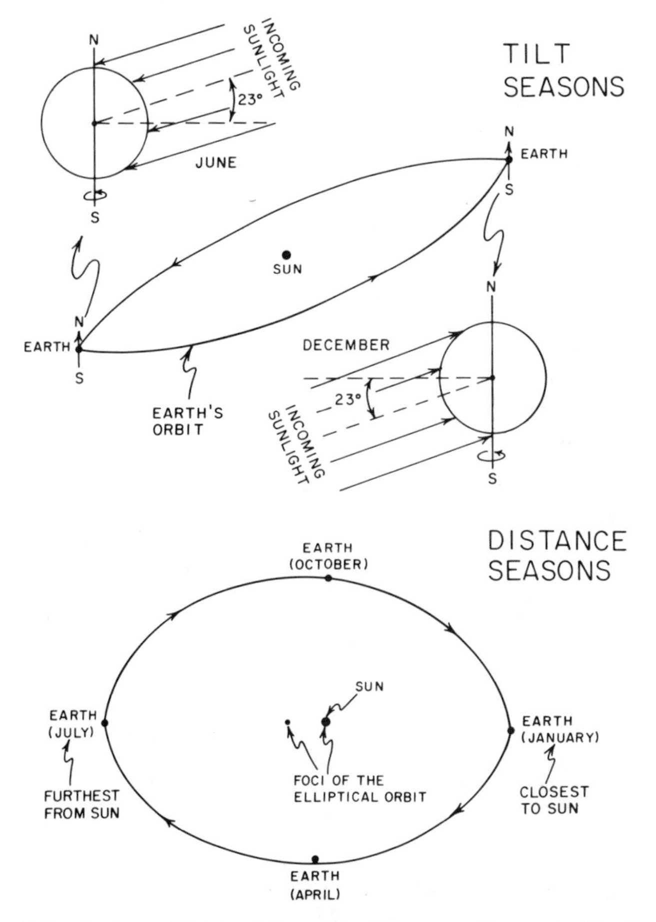

Figure 7-9. Seasonality and its cyclic changes: Shown in this diagram are the elements of the Earth's motion that influence the seasonal contrast in the amount of radiation received at any point on the globe. The primary cause of seasons is the tilt of the Earth's axis with respect to its orbit about the Sun. As shown in the upper panel, this tilt leads to higher illumination of the northern hemisphere during the month of June and to higher illumination of the southern hemisphere during the month of December. Our calendar is set so that June 21st and December 21st correspond to the days on which maximum radiation falls on the northern hemisphere and on the southern hemisphere, respectively. A second contribution to seasonality stems from the fact that the Earth's orbit is elliptical. Because of this the Earth–Sun distance changes during the course of the year. As shown in the lower panel, currently the Earth is closest to the Sun early in January and farthest from the Sun early in July. Thus the Earth as a whole receives less sunlight during July than during January.

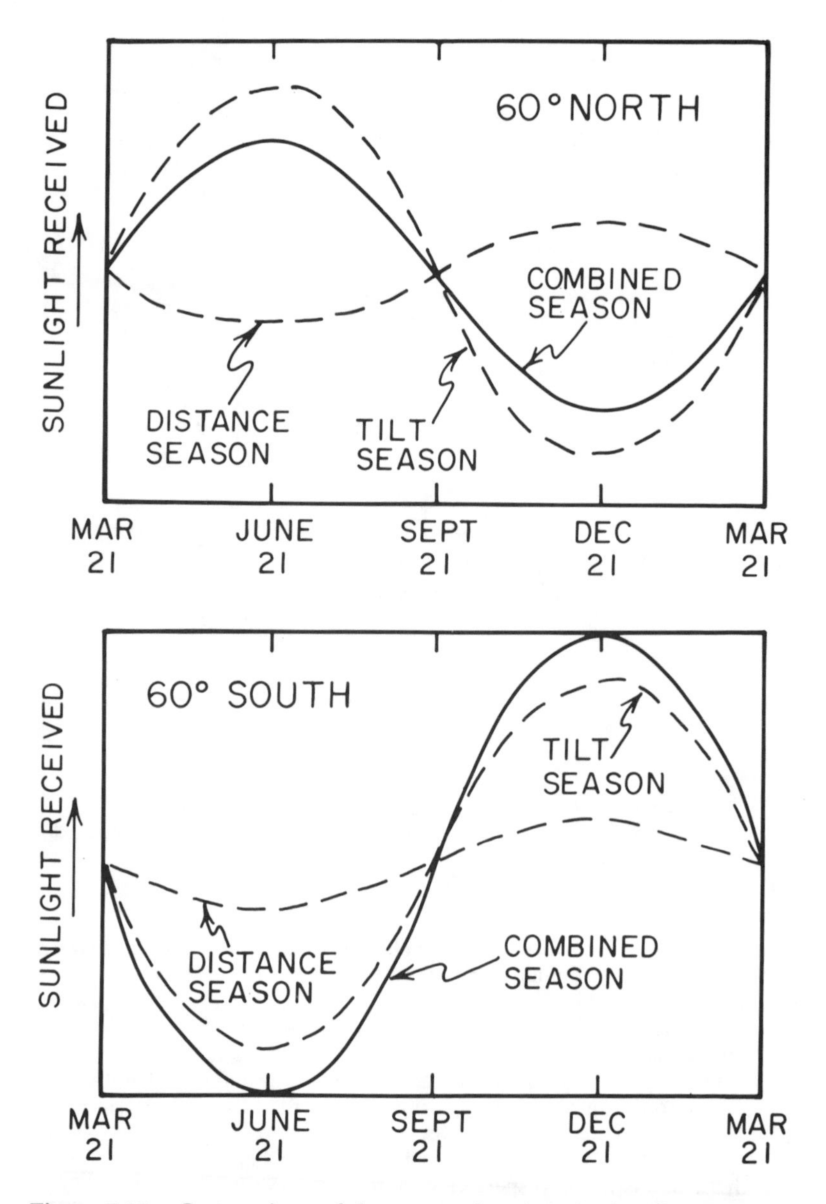

Figure 7-10. Comparison of the seasonal cycles of solar illumination at 60ºN and 60ºS: In the southern hemisphere the tilt and distance seasons currently reinforce one another. In the northern hemisphere they currently oppose one another.

through that part of its orbit that is located farthest from the Sun. By contrast, the Earth points its southern hemisphere toward the Sun when it is traversing that part of its orbit that is closest to the Sun. Hence, the two causes of seasonal contrast (tilt and distance) reinforce one another in the southern hemisphere.

Our interest in all this arises because the situation changes with time. The reason for this is that the Earth precesses as does a spinning top. Both the Earth and the top precess for the same reason: their spin-axis tilt. The leaning top precesses to keep from falling. The leaning Earth precesses to keep its equatorial bulge from becoming aligned with its orbit (that is, pointing directly at the Sun).

Just as a top turns on its axis many more times per minute than it precesses, Earth turns on its axis many more times than it precesses. It takes Earth 8,760,000 days (24,000 years) to complete one precessional cycle.

The importance of precession to seasonality is that it causes a progressive change in the point in the orbit at which the Earth is located at the time its northern hemisphere faces toward the Sun. As shown in Figure 7-11, 12,000 years ago the northern hemisphere faced the Sun as the Earth traversed that part of its orbit nearest to the Sun. Hence, at that time the tilt and distance seasons reinforced one another in the northern hemisphere. Thus, the precession of the Earth causes the contrast between the seasons to change in a cyclic manner.

In addition to its precession, the Earth's orbit is subject to two other cyclic changes. Both have to do with the gravitational tug of the other planets, in particular, that by the large and relatively closeby planet Jupiter. These tugs cause the shape of the Earth's orbit to change with time. At some times it has been even more eccentric than it is today. At other times it has been less eccentric than today. The more eccentric the orbit, the stronger the distance seasonality. These planetary tugs also lead to changes in the Earth's tilt. The greater the tilt, the greater the seasonal contrast; the smaller the tilt, the smaller the seasonal contrast. The time histories of both eccentricity and the tilt of the orbit can be calculated very precisely from a knowledge of the masses

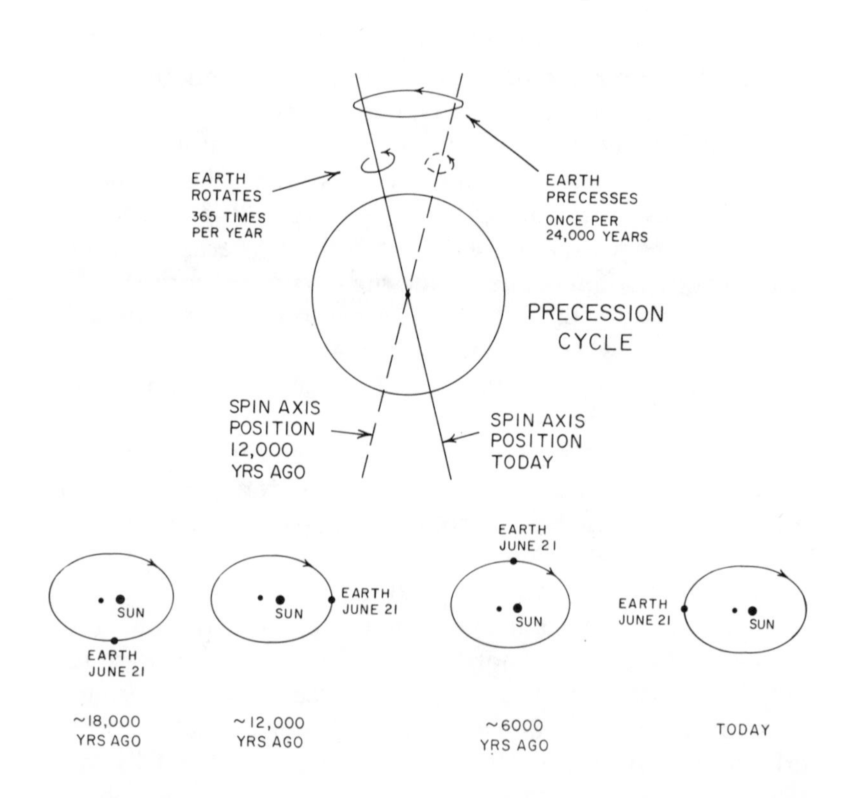

Figure 7-11. Like that of a top, the Earth's spin axis precesses: It takes about 24,000 years for one complete precession cycle. The effect of this motion is to change the point on the Earth's orbit where maximum northern-hemisphere illumination occurs. Today it occurs at the "long" end of the ellipse, reducing full summer illumination in the northern hemisphere a bit. One-half precession cycle ago (i.e., 12,000 years ago) it occurred at the "short" end of the ellipse, increasing full summer illumination a bit. The precession of the Earth's axis is related to the Sun's pull on the equatorial bulge produced by the Earth's rotation. Just as the Earth's gravity seeks to tip over a spinning top, the Sun's gravity seeks to remove the tilt of the Earth's axis. As does a top, the Earth compensates for this pull by precessing.

of the planets and their current orbits. The results of these reconstructions are shown in Figure 7-12. As can be seen, the tilt cycle is quite regular, the highs being spaced at roughly 40,000-year intervals. While the eccentricity changes have a more complex pattern, the highs are spaced at intervals of about 100,000 year.

The Earth's precession, the cyclic changes in its tilt, and the cyclic changes in the shape of its orbit combine to yield a complex history in seasonal contrast. This record is different for different latitudes. The reason is that the influence of tilt changes is prominent only at high latitudes, while the influence of distance changes is equal at all latitudes. In Figure 7-13 is shown the average daily radiation received during the month of July at 65°N (the latitude around which the ice caps of glacial times were centered). The influence of the 24,000-year precession cycle and of the 100,000-year eccentricity cycle are clearly seen. While not immediately obvious, that of the 40,000-year tilt cycle is also present. At this latitude the amount of radiation received during July has varied from a high of 1025 cal per cm^2 per day to a low of 815 cal per cm^2 per day!

The ice-volume record is compared with the summer sunshine record for 65°N in Figure 7-13. While the two curves are far from identical, they do show some intriguing similarities. For example, the rapid disappearances of the great ice sheets (that is, times when the ^{18}O to ^{16}O ratios for benthic forams dropped sharply) occur at times when the summers at high latitudes in the northern hemisphere were unusually warm. Thus, the great ice caps of glacial times appear to have been melted away by the heat of the unusually warm summers.

Further, the smaller wiggles in the ice-volume record nicely track the changes in summer sunshine. John Imbrie of Brown University used a clever mathematical scheme to demonstrate this. He was able to extract from the $^{18}O/^{16}O$ record the variability connected with the precession frequency and the variability connected with the tilt frequency. The dramatic results of this exercise are reproduced in Figure 7-14. As can be seen, the amplitude of the changes in the $^{18}O/^{16}O$ record associated with the precession frequency is as expected: small during times of low eccentricity and large during times of large orbital eccentricity. On the other hand, the amplitude of the changes in the $^{18}O/^{16}O$ record associated with the tilt frequency is as expected: more nearly constant with time. In this way Imbrie makes a compelling case that Earth climate is influenced by the Earth's orbital cycles!

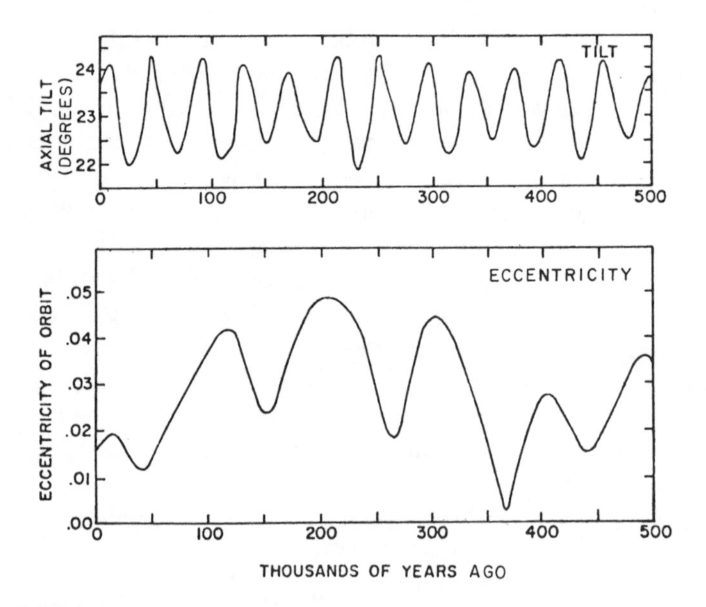

Figure 7-12. The tilt of the Earth's spin axis and the eccentricity of its orbit as a function of time: These records are reconstructed from calculations based on the law of gravitation and the present-day orbits and masses of the planets.

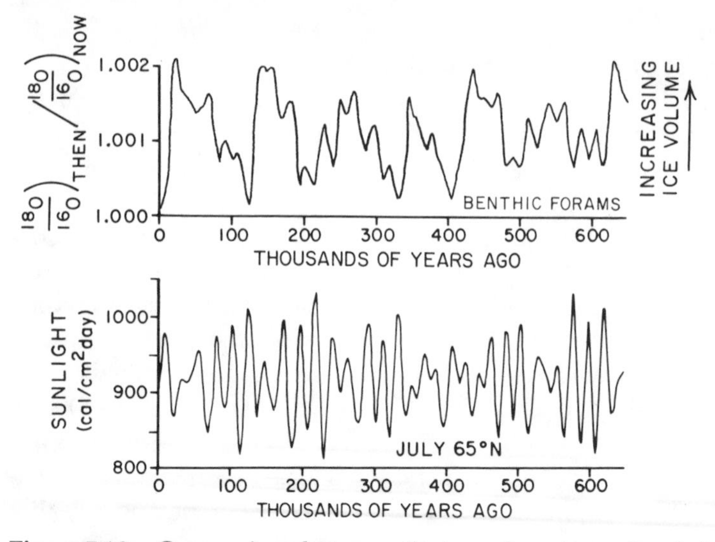

Figure 7-13. Comparison between the ice-volume record and the July solar-radiation record for 65°N latitude.

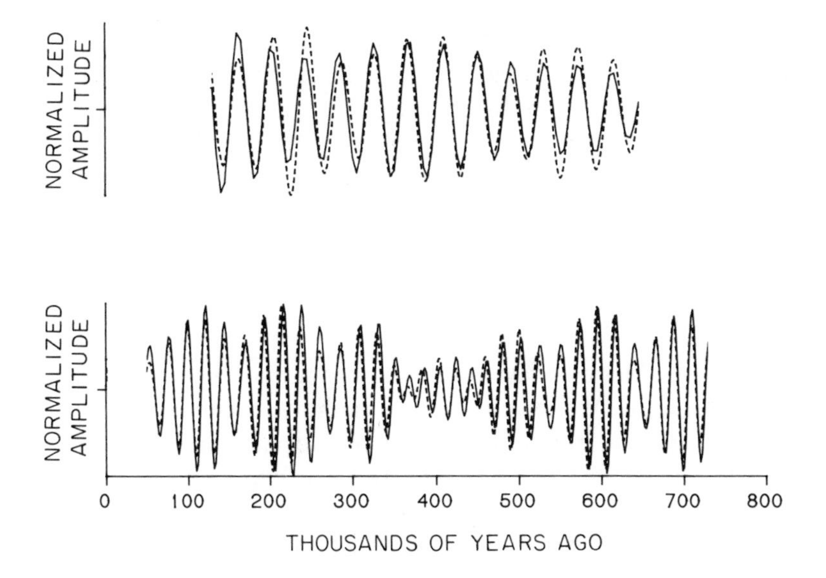

Figure 7-14. Verification that Earth orbit changes produce climate changes: In the upper diagram the amplitude history of tilt-induced seasonality changes (dashed curve) is compared with the amplitude history of the variability in the $^{18}O/^{16}O$ record associated with the tilt frequency (solid curve). In the lower diagram the amplitude history of the precession-induced seasonality changes (dashed curve) is compared with the amplitude history of the variability in the $^{18}O/^{16}O$ record associated with the precession frequency (solid curve). The near-perfect match provides dramatic evidence that Earth climate senses changes in seasonality. The comparison does not extend to the present because the mathematical technique used by Imbrie to isolate the variability in the ^{18}O record associated with a given frequency does not work near the "ends" of the record.

While these correspondences between the summer-sunshine record and the ice-volume record have convinced most geophysicists that changes in the Earth's orbital characteristics somehow drive glacial cycles, the exact connection remains vague. The most popular proposal is that ice caps respond more to changes in summer radiation than to changes in winter radiation. The reasoning is that while the warm summers corresponding to times of higher seasonal contrast take a heavy toll by melting the ice caps, the colder winters that accompany these warmer summers do not generate a compensating amount of extra snowfall. Thus, ice caps suffer during periods of enhanced seasonal contrast (because of the warmer summers) and prosper during periods of reduced seasonal contrast (because of the cooler summers). There are indications that the orbital changes also lead to changes in the patterns of ocean circulation and to changes in atmospheric CO_2 content.

Summary

Unlike the Moon and Mercury, the Earth was able to hold onto its gases. Unlike Venus, the Earth somehow avoided a runaway greenhouse catastrophe. A natural control mechanism involving the carbon cycle on the Earth's surface has apparently prevented our planet from becoming an ice-covered ball. While the Earth has steered a relatively constant temperature course which has prevented its water from becoming all ice or all steam, its temperature has fluctuated considerably. The most dramatic evidence for these fluctuations comes from the record of glaciations which of late have plagued our planet roughly once each 100,000 years. The advances and retreats of these large ice sheets appear to have been driven by cyclic small changes in the Earth's orbital characteristics. If so, a planet's climate is sensitive to minor details of the sizes and orbits of neighboring planets.

Supplementary Readings

Ice Ages: Solving the Mystery, *by John Imbrie and Katherine Palmer Imbrie, 1979, Enslow Publishers.*

A very readable history of thinking as to the role of orbital change as the pacemaker for glacial cycles.

Atmospheres, *by Richard M. Goody and James C.G. Walker, 1972, Prentice-Hall, Inc.*

The basics of the operation of the atmospheric heat engine.

The Chemical Evolution of the Atmosphere and Oceans, *by Heinrich D. Holland, 1984, Princeton University Press.*

An in-depth treatment of evidence regarding past compositions of the ocean and atmosphere.

Pyrite Crystals

Chapter Eight

Storing Up Resources:

The Ingredients of a Civilization

The Earth is a chemical factory whose output is elemental concentrates of remarkable purity. It is the availability of these concentrates that made possible the incredibly rapid evolution of our industrial civilization. The needed energy is obtained by reacting hydrocarbon concentrates which are found in sedimentary reservoirs with the oxygen gas concentrate that makes up 20 percent of our atmosphere. The ingredients for cement and steel can be bulldozed from massive sedimentary deposits. Precious metals and gemstones, while not so abundant, are nevertheless present in concentrated form, awaiting discovery by prospectors.

Most of these chemical wonders are the product of the Earth's hydrologic and life cycles. Rain falling on the continents dissolves substance from the soils and rocks through which it percolates. These salts are carried to the sea where they are precipitated eventually as chemical concentrates. Waters that percolate deep into the Earth's crust lose their dissolved oxygen and also become quite hot. These hydrothermal waters are capable of dissolving trace metallic elements from the rocks through which they pass. When the waters return to the surface, they reprecipitate these metals as ores.

The residues of plant matter contained in sediments, if buried deep enough, are heated and partially converted into gaseous and liquid hydrocarbons. These hydrocarbons are carried from their birthplaces by the circulating waters to geologic reservoirs where they form petroleum and natural-gas concentrates.

Energy from the Sun drives the cycle of evaporation and rainfall and also the cycle of photosynthesis and respiration. Energy from the decay of radioisotopes within the Earth not only drives the mantle's convection cells, but it also heats the deeply circulating waters. Mankind now reaps the harvest from a half billion years of these slow but steady Earth processes.

Introduction

If a planet is to be capable of supporting a technological civilization it must be endowed with some special chemical concentrates. These concentrates are the raw material from which the civilization is built. Planet Earth prepared well for the advent of mankind. Stored within our reach are the chemical concentrates needed for energy generation, manufacturing, and agriculture.

Of the chemical concentrates available to Earth's inhabitants, the most important is O_2 gas. Not only do we need it to burn fuels, but it was absolutely essential to the evolution of complex life forms. While bacteria can "burn" organic material in environments free of oxygen, animals cannot. The difference is that bacteria are unicellular. They absorb both their food and the oxidant used to burn this food directly through their cell walls. Animals have a transport system (blood) which carries the fuel (dissolved sugars) and the oxidant (oxygen gas bound to hemoglobin) to the individual cells. This distribution system allows animals to have cells specialized for seeing, smelling, feeling, remembering, and other functions. On our planet at least, life has not found another oxidant capable of being transported in high enough concentration to permit the development of multicellular organisms. Thus, one requirement for intelligent life appears to be an O_2-bearing atmosphere.

Another basic ingredient is fuel. Civilization reached a level of complexity some two hundred years ago where the energy obtained from plant matter was not adequate.Even the addition of muscle power provided by wind- and water-driven mills was not enough. Man was forced to look elsewhere. While most of the uneaten residues of plant and animal matter are dispersed in trace quantities in sediments, a small fraction became concentrated into nearly pure form as coal, oil, and natural gas. This pure stuff now supplies most of our energy. We are now hooked on these fuels. In industrial countries each person benefits from about ten times more fossil-fuel energy than he produces within his own body. In the brief interval of a few hundred years we will largely burn up a supply of these concentrates built up by nature over the last half billion years.

In addition to oxygen gas and fossil fuels, humans depend on a host of other resources. As if it foresaw this need, planet Earth provided virtually every one of them in nearly pure form. In so doing, it made the task of creating our complex civilization remarkably easy. Even though they are often rare and hidden from our immediate view, nature's amazing stockpiles eventually were discovered.

The task for this chapter will be to understand how our planet generated the substances required for life: the molecular oxygen necessary for the evolution of animals (and for the burning of fossil fuels); the fossil fuels we needed to carry us through the first phase of our technological development; the fertilizer we need to grow food for the billions of humans; and the ores required to make the innumerable products demanded by the Earth's population.

Atmospheric Oxygen

Our atmosphere consists of a mixture of about four parts N_2 and one part O_2*. The N_2 is the major reservoir for the element nitrogen in the Earth. Since N_2 is a very stable molecule, few scientific eyebrows are raised by its presence in our atmosphere. Rather, it is the atmosphere's O_2 that raises eyebrows!

Were this O_2 to be exposed to the immense amounts of H_2 gas initially present in the nebula, it would have reacted to form water. A chemist would term the conditions that prevailed in the nebula as highly "reducing." In such situations there are more atoms willing to donate electrons to their neighbors than there are atoms willing to accept electrons from their neighbors. One important consequence of the willingness of hydrogen to donate electrons was the conversion of iron oxide to iron metal via the reaction

$$FeO + H_2 \rightarrow Fe + H_2O$$

*As shown in Table 2 of Chapter 7, the atmosphere also contains some water vapor and 1 percent argon (almost all of it ^{40}Ar produced by the decay of the Earth's ^{40}K) and traces of CO_2, H_2, CH_4, N_2O, and of the other noble gases.

In iron oxide, the iron atoms have two fewer electrons than are required for electrical neutrality. These electrons orbit a neighboring oxygen atom, giving it two more electrons than required for neutrality. The solid is held together by the attraction exerted by the positively charged iron ions on the negatively charged oxygen ions (see Figure 8-1). In the chemical reaction shown above, the iron atoms regain their missing electrons from hydrogen. The positively charged hydrogen atoms created in this way become attached to the negatively charged oxygens, creating water molecules.

In the case of the Earth's interior, the reducing conditions are the result of the presence of a massive iron core (which, of course, in turn stems from the dominance of H_2 in the nebula). Were the Earth's atmospheric oxygen to be placed in contact with this iron it would be destroyed via the reaction

$$2Fe + O_2 \rightarrow 2FeO$$

In this reaction iron atoms donate electrons and oxygen atoms accept them.

So we live on a planet whose surface is dominated by the electron-accepting desire of O_2 gas and whose core is dominated by the electron-donating desire of iron metal. If a bit of the Earth's core were placed in contact with the atmosphere the metallic iron atoms would give electrons to molecules of oxygen in the atmosphere. The metallic iron would be converted to rust. Similarly, if a bit of air were injected into the Earth's core, the oxygen would be consumed to form iron oxide. The important thing is that if all the materials making up the Earth were to be stirred together, the core's iron would overwhelm the atmosphere's oxygen. As there are more iron atoms by orders of magnitude in the Earth's core than oxygen gas molecules in its atmosphere, only a tiny bit of the core would be converted; iron oxide would form before the oxygen gas was gone. Thus, in accord with its nebular heritage, the Earth is a highly reducing environment. Somehow in the course of geologic time the atmosphere countered this tendency and became oxidizing. A neat trick!

These oxidizing conditions set by the atmosphere are confined to the outermost skin of the Earth. No matter where we penetrate

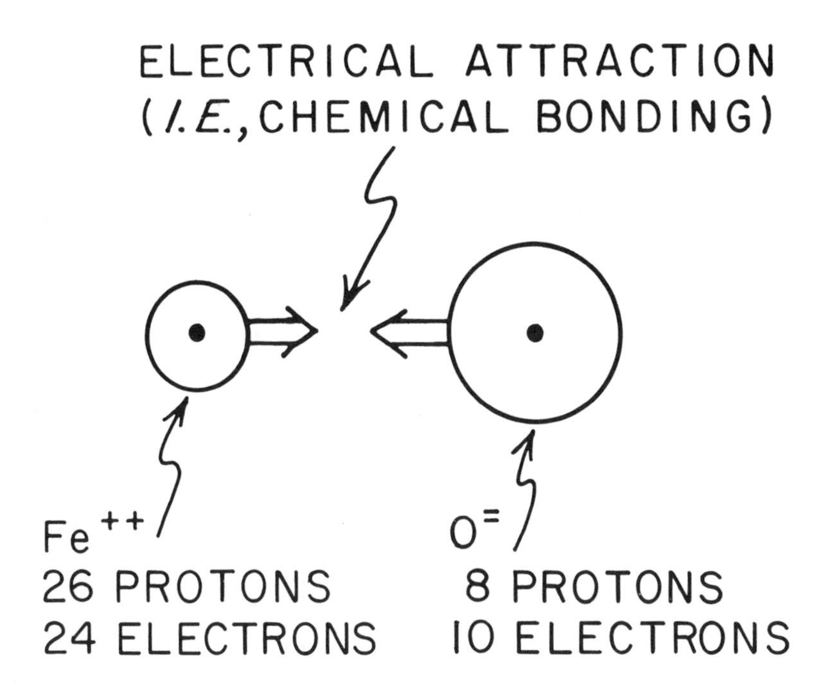

Figure 8-1. Chemical bonding: In O_2 form each oxygen atom has a number of electrons equal to its number of protons. Since oxygen atoms prefer to have two extra electrons, these O_2 molecules can act as electron acceptors (known as oxidizing agents). In iron metal form each iron atom has a number of electrons equal to its number of protons. Since iron atoms prefer to have two less electrons than this neutrality number, metallic iron atoms can act as electron donors (known as reducing agents). Thus, when O_2 gas encounters iron metal an electron transfer occurs. The iron metal gives up electrons to the O_2 molecules. The Fe^{++} and $O^{=}$ ions made in this way are strongly attracted to one another. They merge into a chemical compound FeO. This compound is bound together by the mutual electrical attraction of its oppositely charged constituent ions.

the Earth's surface, we soon leave the oxidized realm and enter the reduced realm. In the Black Sea this transition occurs within the water column itself; the deep waters are free of oxygen. In an algae-choked pond it occurs only a few millimeters below the sediment surface. In the open ocean the transition occurs hundreds of meters deep in the sediments. At no place do oxidizing conditions extend beyond a kilometer or two into the solid Earth.

Because of its heritage, the Earth's atmosphere must initially have been O_2-free. During the course of geologic time, the Earth's surface environment was somehow transformed from reducing to oxidizing. This switch-over was a very important event in the evolution of life. Without O_2 life was doomed to remain quite simple; each cell was forced to be chemically self-sufficient.Only when O_2 became available was there a means of transporting chemical energy to specialized cells. O_2 and organic molecules travel together through blood vessels to the site where energy is needed. Here an enzyme triggers combustion. Given an atmosphere with O_2, animal life took off, evolving everything from mosquitoes to dinosaurs!

One has only to probe the mud at the bottom of a pond to realize that not even worms can live in environments lacking O_2. These chemically reducing domains are still ruled by simple bacteria.

We do not know for sure how the transformation from a reducing to an oxidizing atmosphere was accomplished. One idea is that O_2 was slowly created by the escape of H_2 from the top of the atmosphere. As discussed in the preceding chapter, water molecules reaching the "top" of the atmosphere are decomposed by packets of the Sun's ultraviolet light.* The light hydrogen atoms freed in this way could evaporate while the heavier oxygens would be left behind. Clearly, the first O_2 to be produced in this way would have been gobbled up promptly by the abundant electron-donating substances initially present at the Earth's surface. With time, however, these donors would have been depleted, allowing O_2 to begin to accumulate in the atmosphere.

Another possibility is that life itself did the job. Algae, one form of early life, produced organic matter and O_2 gas. The organic material made by these plants was in part consumed by bacteria.

---

*We made the point in the preceding chapter that only a small fraction of the Earth's hydrogen had been lost in this way. However, only a small fraction of the hydrogen need be lost to create the O_2 we have. Currently there are about a thousand water molecules in the ocean for every O_2 molecule in the atmosphere.

Some of the plant matter, however, went uneaten and was buried in sediments. For each unit of carbon buried, a unit of O_2 remained in the atmosphere. Thus began the segregation of reducing compounds (the buried organic carbon) and oxidizing compounds (the O_2). As in the case of loss of hydrogen to outer space, this early oxygen must have been consumed by the hosts of reducing compounds present in the atmosphere, in the ocean, and in soils. Finally, after several billions of years, O_2 displaced the reducing gases H_2, CH_4, and CO from the atmosphere and other reducing compounds from the ocean and from the soils. Once in command, O_2 held the upper hand and retained its dominance over the persistent invasion of reducing compounds from just beneath the planet's surface.

While not being able to evaluate the relative importance of these two processes, we do have some idea of *when* the Earth's surface conditions left the reducing realm and entered the oxidizing realm. As for planet formation, the "when" is more easily pinned down than the "how." The fossil record puts the latest possible date on the transition. The age of the remains of the earliest preserved fossil of O_2-utilizing animals fixes this limit. The animals turn out to be worms (see Figure 8-2). The oldest worm fossils found so far are about 0.7 billion years old. Thus, it may have been as late as 3.8 billion years after its formation that the Earth's surface turned the corner and became sufficiently oxidized to permit animal life to evolve.

Another type of evidence comes from what geologists call sedimentary iron formations. While such deposits are common in sedimentary sequences greater than 1.9 billion years in age, they are rare in sediment sequences in the age range 1.9 to 0.7 billion years and absent in sequences more recent than this. To understand their significance we must digress to learn a bit about sediments. In so doing we shall also encounter our first examples of the process whereby nature produces chemical concentrates.

Figure 8-2. The oldest known animals: The fossils from the late Precambrian Ediacara formation of Australia are preserved as impressions in sandstone. Shown here is the imprint of a worm. (Courtesy R. Sprigg).

Chemical Precipitates from the Sea

The products of continental erosion can be divided into two major categories: dissolved and detrital. The detrital products include everything from the boulders and cobbles found at the bottoms of rushing mountain streams to the very tiny mineral grains that make up the muds found at the bottom of the ocean. In this regard ocean sediments are chemical grab bags. As the ocean floor receives mineral grains from a host of rivers and from dust blown off the continents, its sediments have a chemical composition nearly matching the average for the entire crust. Geologists term sediments dominated by this fine detritus as shales.

By contrast, the ions dissolved from continental soils form sediments that often consist of a single mineral.For example, we find large regions of the ocean floor blanketed with sediments that are more than 75 percent the mineral calcite and other large areas that are more than 50 percent the mineral opal (see Figure 8-3). The calcite ($CaCO_3$) and opal (SiO_2) in marine sediments is produced by marine plankton (see Figure 8-4).

The great enrichment of opal in sediments found beneath the Antarctic Ocean, beneath the equatorial Pacific Ocean, and beneath the eastern tropical Atlantic is the result of unusually high plant growth rates in the overlying surface waters. The plants in these waters are mainly diatoms. The high growth rates are sustained by the upwelling of subsurface waters. Plant growth in the ocean is generally held back by the lack of sufficient dissolved nitrate and phosphate (that is, fertilizer). These critical ingredients are rapidly removed from surface waters as falling organic debris (dead organisms, fecal matter), leaving most of the ocean's surface waters nutrient-poor. Animals and bacteria living in deep waters oxidize the falling organic debris and return the bound N and P to dissolved form (NO_3 and PO_4). Thus, in places where these deep waters are drawn back to the surface, plant life thrives. Such "upwelling" occurs in the tropics and in the Antarctic. In these areas the rain rate of opal far exceeds that of continent-derived detritus, producing sediment very rich in opal.

Unlike diatoms, which live primarily in areas of upwelling, organisms that produce calcitic hard parts live in the remaining portions of the surface ocean where fertilizer is in short supply. Although the calcitic remains of these organisms fall to the sea floor virtually everywhere, sediments rich in calcite are found capping the midocean ridges and plateaus much as snow caps high mountains. The reason is that the deepest waters of the ocean are corrosive toward calcite. Calcite falling to these deep parts of the sea dissolves. The calcite houses made by

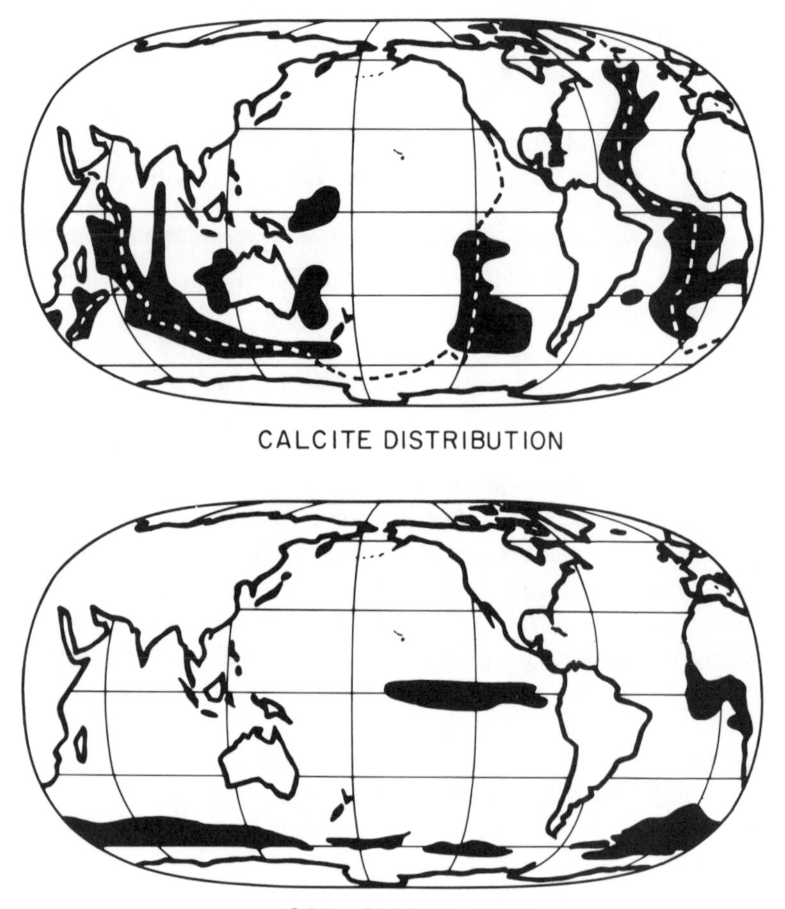

Figure 8-3. The distribution of calcite and opal sediments on today's sea floor: Sediments containing more than 75 percent calcite are shown by the black areas. These areas largely correspond to the deep-ocean mountain system associated with the rising limbs of mantle convection cells. The rift along which new basalts form is shown by the dashed line. Sediments containing more than 50 percent opal are shown by the black areas. These areas correspond to zones where deep water upwells to the surface.

marine plankton are preserved only on sea-floor "mountains" which reach above corrosive bottom waters. Away from the immediate margins of the continents the rain rate onto the sea floor

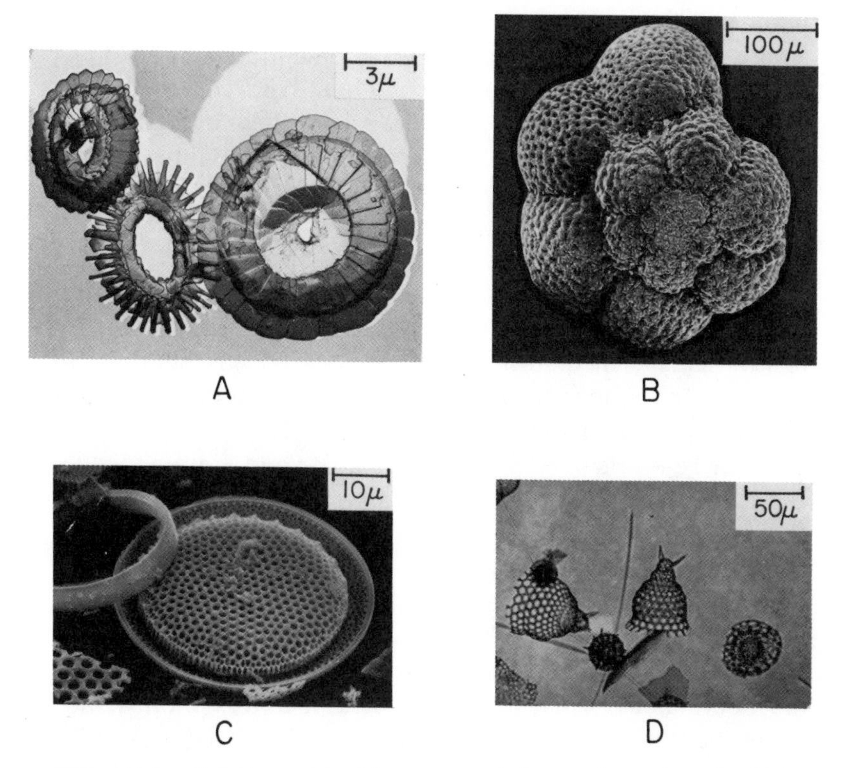

Figure 8-4. Calcite and opal houses made by marine microplankton: (A) Calcite cages which surround plants called coccolithophorida. (B) Calcite shell formed by foraminifera. (C) Opaline pillbox housing a diatom. (D) Opaline cages inhabited by radiolarians. These pictures were taken using a microscope; there are 10,000 microns in a centimeter.

of calcite exceeds that of detritus. Thus, in those regions of open ocean where calcite does not redissolve it dominates the sediment.

For those areas of the sea floor where neither opal nor calcite accumulates the sediment is a mixture of fine detritus brought to the sea by rivers and winds.

Two other important chemical precipitates are found in sediments formed in the recent past. They are gypsum ($CaSO_4 \cdot H_2O$) and halite (NaCl). They differ from the calcite and opal found in ocean sediments in an important way. They are not formed by organisms. Rather, they precipitate spontaneously from seawater. As today's ocean is undersaturated with respect to both of these minerals, they are not found in recent sediments. It is thought that geologic strata made up of these minerals were formed in isolated arms of the sea (the Persian Gulf and Red Sea come closest to this situation in today's ocean). These arms acted as gigantic evaporation pans.

In sediments formed more than 2 billion years ago, another spontaneous chemical precipitate is commonly found. It is iron oxide. Not only don't we find any evidence for similar deposits in sediments formed during recent geologic epochs, but they are not expected. The reason is that O_2 reacts with the Fe^{++} ions dissolved from rocks, changing them to the Fe^{+++} form. It turns out that Fe^{+++} is very insoluble in water and is immediately reprecipitated as coatings on the fine mineral debris of which shales are made. To form sediments dominated by a single mineral the chemical ingredients must be quite soluble in seawater. Only in this way can they be transported to those special environments where massive precipitates can accumulate. The high concentration of oxygen in today's atmosphere dictates that iron be present in waters in only trace amounts. Hence, sediments dominated by iron no longer form.

However, if at some time in the past conditions existed where Fe^{++} instead of Fe^{+++} were the stable chemical form of iron in natural waters, things would have been different. Fe^{++} is quite soluble and could be transported in large quantities to the arms of the ocean that acted as evaporation pans. Geologists surmise that at the time the so-called iron formations were deposited, O_2 had not yet become an important constituent of the atmosphere. Under these conditions Fe^{++} would have been the stable form of iron in the Earth's surface waters. If this assessment is correct, then the age of the last of the great iron formations sets the earliest possible time for the transition from reducing to oxidizing conditions in the atmosphere. This limit turns out to be about 1.9 billion years ago. Thus, as shown in Figure 8-5, the transition from reducing to oxidizing conditions appears to have taken place between 1.9 billion (the limit placed by sedimentary iron formations) and 0.7 billion years ago (the limit set by the appearance of worms).

If this assessment is correct (it gets about a 7 on our scale of probability), then about 3 billion years of Earth history passed before the mechanisms producing oxygen could generate enough of it to overcome chemical species capable of destroying it. If life itself was mainly responsible, then the bootstrap operation proved a very lengthy one. Hence, the creation of an O_2-bearing atmosphere is a task which life on other seemingly suitable planets may not have been able to achieve. On the other hand, if dissociation of water molecules high in the atmosphere was mainly responsible, then the host of characteristics that govern the transport of water molecules to the uppermost atmosphere, the breakup of these molecules, and the escape of the dissociated hydrogen atoms are critical to life's success.

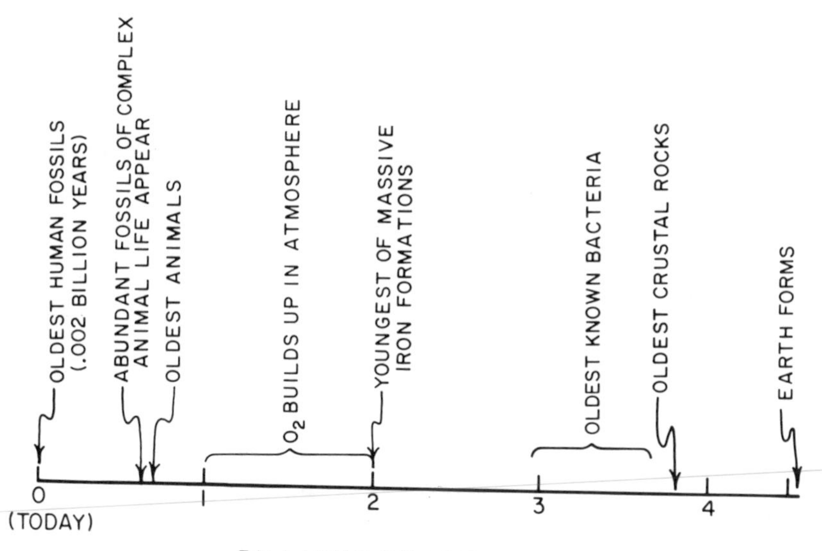

Figure 8-5. Important events in the evolution of life: Bacterial remains are found in the Earth's oldest sediments. Abundant animal remains are found during only the last one-sixth of Earth history. Their evolution had to await the buildup of O_2 in the atmosphere. Complex shell-bearing organisms appeared suddenly about six-tenths of a billion years ago. Man did not appear until very, very recently.

Fossil Fuel

As discussed in Chapter Seven, most of the Earth's carbon is present in limestone (as $CaCO_3$) and in shales (as kerogen). Kerogen is the term used by petroleum geologists for the organic residues dispersed among the mineral grains in sediments. Kerogen is a potential fuel. If it could be concentrated from shales, it could be burned to provide energy:

$$\text{Kerogen} + O_2 \rightarrow CO_2 + H_2O$$

Were this possible, our energy supply would be eternally secured. Unfortunately, this is not possible. The reason is that kerogen rarely makes up more than 1 percent of the sediment. Thus, even if we could make it burn, for every ton of kerogen fuel we would have to feed 100 tons of shale into the furnace and then discard 99 tons of ash! In any case, shale won't burn. Further, there is no economically feasible way to separate the kerogen from the mineral grains. Hence, these residues of eons of life have no direct value to man. Nor will they ever.

Nevertheless, petroleum geologists are intensely interested in kerogen. The reason is that kerogen is the birthplace of oil! Contained in these organic residues is fatty matter. As a sediment becomes buried beneath its more recent counterparts, it is subjected to ever higher Earth temperatures. The kerogen is cooked. The cooking converts some of the fatty matter into liquid hydrocarbons—oil. While much of what is produced remains in the source shale, some of these hydrocarbons migrate through the pores in the rock. In some cases the routes of migration lead to natural traps (see Figure 8-6). From man's point of view the best traps are in porous sandstone (ancient beaches, dunes, river beds, and the like) and porous limestones (ancient coral reefs). Unlike shales, which are very fine-grained and hence quite impermeable, these rocks often are sufficiently permeable that we have only to drill down and pump out pure liquid hydrocarbon (so-called crude oil). Unfortunately for mankind, only a small fraction of kerogen is fatty matter and only a small fraction of this fatty matter has been converted to hydrocarbons, and only a small fraction of these hydrocarbons have moved from their

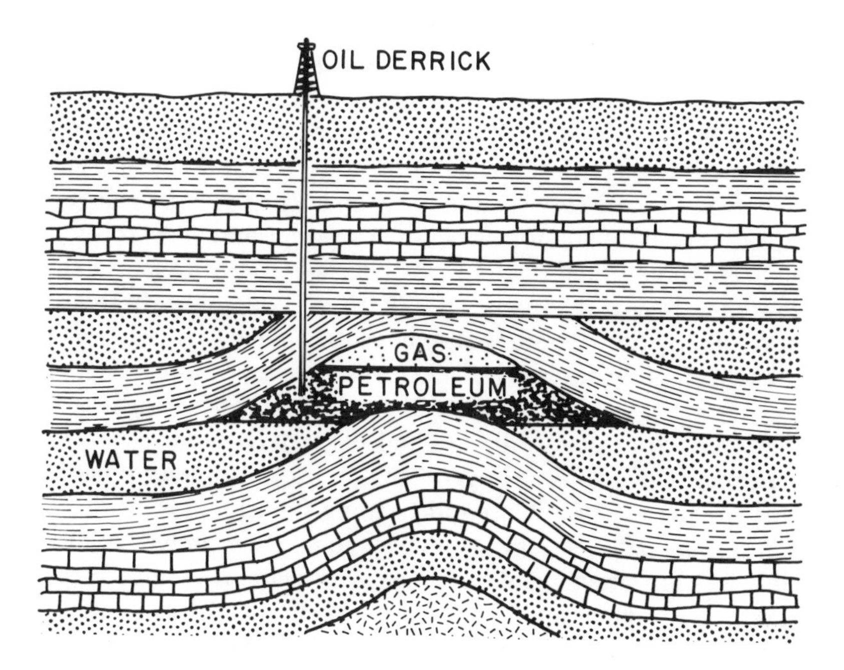

Figure 8-6. A typical petroleum trap: Oil and methane gas are carried by circulating waters from their birthplaces in shales into neighboring porous sandstones. If the sandstone has been bent into an arch, as is shown here, then the oil will float above the water and fill the arch. If methane is present it will form a gas pocket above the oil. These hydrocarbons are trapped in the sandstone by the impermeable overlying shale.

birthplaces to traps suitable for the driller's straws. We have already sucked up about 15 percent of the oil nature has stored up for our use. We will likely recover much of the remainder during the next hundred years. Needless to say, oil is not a renewable resource! Only after several hundred millions of years could a supply be regenerated comparable to what we have already tapped.

The natural gas we burn in our stoves and furnaces consists primarily of methane. This gas also is created when the organic residues in sediments are heated; the heating must be more intense, however. Like oil, once formed, methane migrates. That captured in sedimentary traps awaits the pipes of man. Because this gas is pressurized by its confinement it does not need to be pumped. Rather, it pushes its way through the pipe, much as air pushes through a leak in a tire.

Coal is the only one of the fossil fuels not to have been concentrated by migration through the pores of sediments. Rather, coal seams are primary sedimentary units. The modern analogue of coal is peat, which forms beneath swamps and bogs. In such environments much of the plant debris goes uneaten and accumulates. Also, relatively little detritus is added to these environments from the surrounding lands. Hence, the sediments consist primarily of plant remains. Once buried, the organic matter is compressed and baked by Earth heat, transforming it progressively from peat to lignite and ultimately to coal.

Humans have become dependent on energy. With our population already over 4 billion and heading for 10 billion or more before stabilizing, we cannot go back to an agrarian way of life. Only by supplementing our muscle power with many times more energy from other sources can we prevent massive starvation. The Earth's fossil fuel reserves will suffice for this purpose for only a hundred years or so. As this resource dwindles, we will either have to learn to convert solar energy into electricity on a large scale or turn to some mode of nuclear-energy generation.

Metal Ores

Fortunately for mankind, O_2 and fossil-fuel concentrates are only two of many entries on the list of Earth's amazing chemical concentrates. We have seen that during the early history of the Earth massive sedimentary iron oxide deposits formed on the planet. Though only a fraction of these deposits survive, they are more than ample to supply the raw material for such products as automobiles, bridges, and rails. The iron in these deposits is relatively easily converted to metallic form. Were we instead forced to separate iron from basalt, all the products made of iron would be far, far more expensive.

We have also seen that sediments of pure gypsum ($CaSO_4 \bullet H_2O$) formed in isolated basins of the sea. Motions of the Earth's plates have raised some of these above sea level where they can be harvested with a bulldozer and converted to economical (and fireproof) wallboard. Likewise, sediments of nearly pure $CaCO_3$ can be scooped up and converted into concrete.

The most amazing of nature's concentrates are the precious metal ores. Gold, whose average earth abundance is only about 10 parts per billion, is found in nuggets! So also is platinum, whose earth abundance is only about 100 parts per billion. The mercury we use in our thermometers is found in ores of cinnabar, a mineral consisting of the elements mercury and sulfur. The average abundance of the element mercury in the Earth is only a few parts per billion. Were we forced to separate these metals from granite or seawater or any other common Earth substance, their cost would be at least a thousand times higher!

These and the host of other metal ores were produced by the percolation of hot waters through the Earth's crust. Cold water seeps into the crust from soils and into cracks in the underlying rock. The deeper it seeps, the warmer it gets. Since warm water is less dense than cold water, this sets up the tendency for the water to move in organized cells. Cold water penetrates along one arm of the cell and warm water is driven back to the surface along the other. Old Faithful, the famous geyser in Yellowstone Park, is part of such a geothermal circulation system.

As these geothermally driven waters move through the crust they dissolve traces of metal from the rocks through which they pass. These metals are carried in solution for some distance. Changes in temperature and oxidation state along these routes eventually cause these metals to reprecipitate. This chemical mill is driven in part by the escaping heat from the radioactive decay of the Earth's U, Th, and K and in part by the great contrast between the highly oxidized conditions at the Earth's surface and the highly reducing conditions in its interior. Like the chromatographic columns used by chemists, Earth rocks are able to selectively concentrate on individual metals!

A dramatic example of such a hydrothermal circulation system was discovered during the 1970s far beneath the sea. Using a deep manned submersible operated by Woods Hole Oceanographic Institution, scientists exploring the great crack in the ocean crust created by the divergence of the underlying mantle convection cells stumbled on an amazing scene. In contrast to the nearly lifeless conditions seen on most of the deep sea floor, certain segments of the rift zone teemed with strange creatures. Clams three times as big as those we make chowder from, white bamboo-like stalks capped with lipstick-like red tips, spider-like crabs crawling through the jungles of stalks—such were the life forms that dazzled the eyes of those lucky observers!

It was quickly realized why these sea-floor oases existed. Springs of hot water discharging from the rift allow food to be produced which maintains these isolated habitats. This food is not made by plants; rather, it is made by bacteria. The synthesis of organic molecules in these bacteria is fueled with energy obtained by combining the H_2S gas dissolved in the emerging hot water with the O_2 gas dissolved in the deep sea waters. The H_2S reflects the reducing conditions of the Earth's interior, while the O_2 reflects the oxidizing conditions of its surface. Using the energy derived from this chemical reaction and HCO_3^-, NO_3^-, and $H_3PO_4^-$ dissolved in the water, these bacteria multiply. As fast as they multiply, they are eaten by the animals of this bizarre community.

The sulfur found in the spring waters (as H_2S) comes from the seawater that seeps into the sea floor on the flanks of the ridge. This water is heated by the hot rock and rises back to the sea in the rift zone. When hot seawater reacts with basalt, the oxidized sulfur ($SO_4^=$) it contains reacts with the reduced iron (Fe^{++}) in the basalt. Under these conditions the iron is oxidized (to Fe^{+++}) and the sulfur is reduced (to $S^=$).

Not all of the reduced sulfur generated in this sea-floor hydrothermal system goes into generating energy for bacteria. It also goes into making sulfide minerals around the orifices of the springs. As do all hot circulating waters, those percolating through the ridge dissolve metals from the rocks through which they pass. The cooling that occurs as the spring water emerges into the sea causes these metals to precipitate as metal sulfides. So these sea-floor explorers also found iron sulfides and copper sulfides in great quantity associated with the exotic sea-floor oasis.

Nature works its chemical wonders in a variety of ways. In the sections that follow we discuss the source of a sought-after gem, of a major reagent in the chemical industry, and of the critical ingredients of plant fertilizer.

Diamonds

Besides being the world's favorite jewel stone, diamonds are the ultimate material for cutting and grinding. No harder material exists; when rubbed against any other material, diamond wins. No other substance transmits visible light so well. This is why diamonds are so beautiful. Light entering the crystal bounces from the many facets created by the jeweler and sends flashes to the observer's eye. Were certain frequencies of the light to be absorbed within the crystal, the emerging light would be colored. But this does not occur.

The remarkable properties of this crystal stem from its very unusual chemistry. Diamond is made exclusively of carbon atoms. They are very tightly bound to one another in the three-dimensional network described in Figure 8-7. Each carbon atom is surrounded by four neighboring carbon atoms; these are arranged at the corners of a tetrahedron. Because carbon atoms are very small, substitutions by other elements so prevalent in most other minerals do no occur. Diamond is chemically pure.

For long mankind has sought the secret of manufacturing these valuable crystals. Finally, after centuries of failure, scientists discovered how to make diamonds. Carbon had to be heated very hot, while at the same time being subjected to extremely high pressures. So high are these pressures that they are duplicated in the Earth only at depths greater than about 140 kilometers (see Figure 8-8). From their research, the General Electric scientists who made this discovery were able to show that this pressure limitation was universal. Nowhere in the solar system could diamonds have grown at lesser pressures. Hence they concluded that the diamonds found on Earth must have formed deep in its interior and then somehow migrated to the Earth's surface.

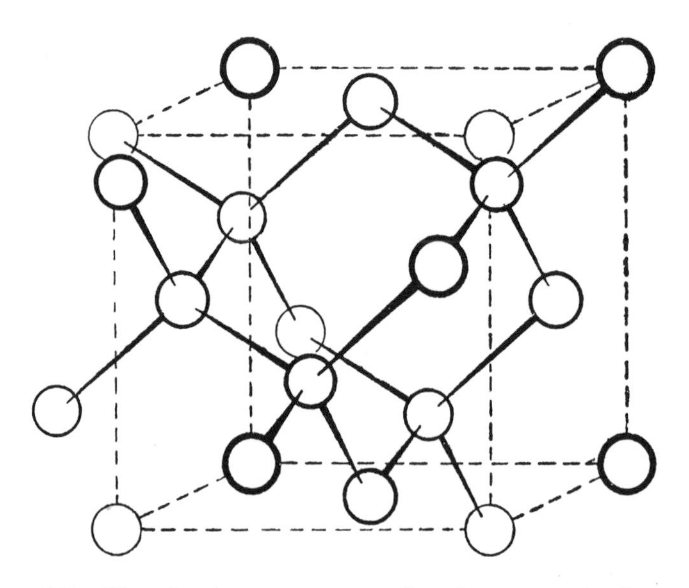

Figure 8-7. The atomic arrangement of carbon atoms in diamond: Each carbon atom has four neighbors to which it is "bonded." These four atoms lie at the corners of an imaginary tetrahedron. On a larger scale these tetrahedra form cubic units. Note that carbon atoms lie at the corners of a cube, on centers of each face of the cube, and in the center of four of the eight smaller cubes making up the big cube.

Geologists found no problem with this conclusion, for they had long noted that diamonds were found only in a very special type of host rock. This rock, called kimberlite, contains an association of minerals akin to that hypothesized for the Earth's mantle. Furthermore, kimberlites are found in pipe-shaped structures which give every appearance of having been punched up in solid form from below. So the geologist's perception was that diamond-bearing host rocks are deformed (but unmelted) pieces of the Earth's mantle which by some unknown mechanism had been thrust up into the crust.

The element carbon makes up about 0.1 percent of the Earth's mass. Yet somehow carbon atoms in the mantle were able to accumulate to form diamond crystals as big as your thumb and in a few rare cases even as big as your fist. Equally amazing is that

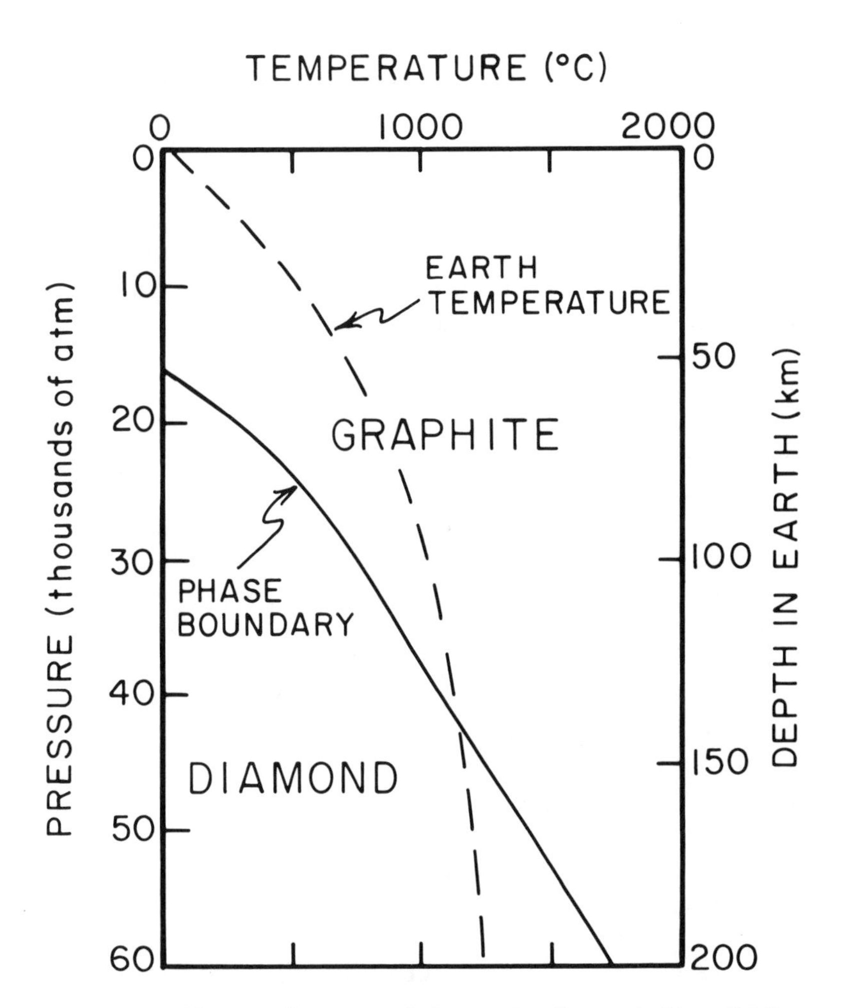

Figure 8-8. The conditions needed to make diamond: The solid line shows the boundary separating the stability fields of carbon in the graphite form and carbon in the diamond form. The dashed line shows an estimate of how temperature in the Earth rises with depth. Since each depth in the Earth has a corresponding pressure associated with it, the laboratory experiments and Earth-temperature curve can be displayed on a single graph. Above about 140 kilometers the Earth-temperature curve lies in the stability field of graphite. At 140-kilometers depth the two curves cross. Below this depth diamonds are stable. This tells us that the diamonds we find in kimberlites could only have grown well below the base of the continental crust.

these crystals were somehow carried from their point of origin to the Earth's surface so quickly that they did not have time to recrystallize to graphite, the low-pressure form of pure carbon.

Native Sulfur

A far less glamorous commodity needed to maintain our industrial society is sulfuric acid. Our chemical plants consume 40 million tons of this substance each year. Most of it is made from elemental sulfur found in nearly pure form within sediments consisting mainly of gypsum and halite. Indeed, the sulfur is thought to have been deposited originally as gypsum.

The natural process that unlocked the sulfur from its very stable gypsum form is one of the many natural chemical miracles that have been performed on our Earth. In its gypsum form, sulfur is economically unsuitable as a sulfuric acid source. The energy required to break the strong chemical bonds that tie it to the neighboring oxygen atoms is too great to permit gypsum to be used as the raw material for acid.

Using isotopic clues, geologists feel that they have solved the mystery of nature's formula for generating deposits of pure sulfur. The first step in this process is the deposition of salts (halite and gypsum) in an isolated arm of the sea. These salts are subsequently buried beneath thousands of feet of limestone and shale. Because of their low density, the halite layers become buoyant and push their way through overlying sediment. As these tongues of salt rise, they deform the sediments through which they push (see Figure 8-9). Ideal petroleum traps are created. Methane gas migrates up the tilted sand horizons until it reaches the impermeable salt beds. Here it pools.

Opportunistic bacteria take advantage of this situation. They have enzymes that allow them to react methane with the sulfate from gypsum. With the energy released during this reaction the bacteria can build the chemical compounds needed by their cells. The products of this chemical reaction are elemental sulfur and the mineral calcite. The calcium from the gypsum becomes the calcium in the calcite. The carbon in the methane becomes the carbon in the calcite ($CaCO_3$), and of course the sulfur in the gypsum is released to elemental form.

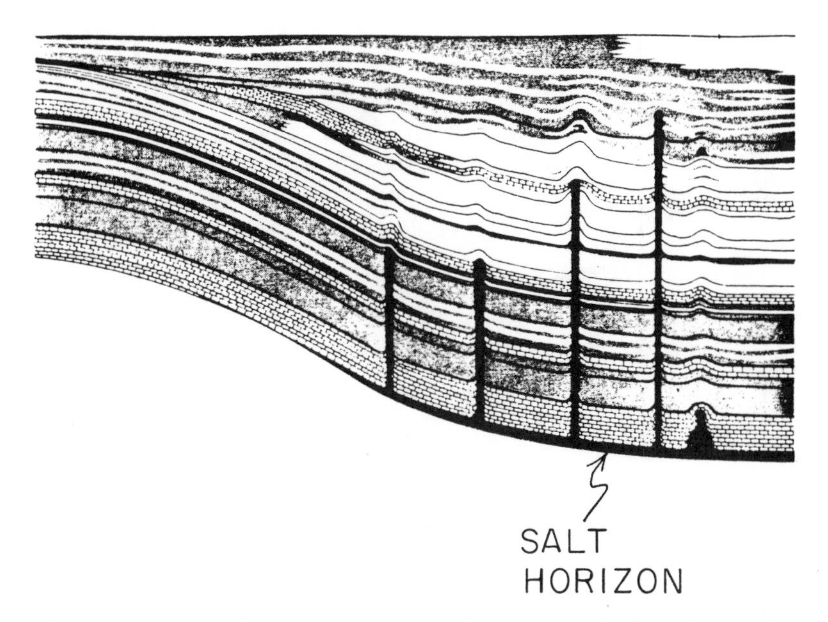

Figure 8-9. Salt domes: Narrow plug-like columns of salt are believed to "flow" upward through the more dense but mechanically weak overlying sediments, forming "salt domes." This geologic section of eastern Louisiana shows known salt domes that have risen through as much as 10,000 meters of overlying sediments.

The reason why geologists are confident that they have figured out nature's mechanism for generating elemental sulfur is that the ratios of the carbon isotopes in the calcite lend strong support to their reconstruction. During photosynthesis plants utilize $^{12}CO_2$ in slight preference to $^{13}CO_2$. Because of this the $^{13}C/^{12}C$ ratio in all plant matter is lower than that for atmospheric or ocean carbon. It is also lower than that in the calcite formed by marine organisms. Thus, when ordinary sediments are subjected to isotopic analysis, the carbon locked up in $CaCO_3$ always has a $^{13}C/^{12}C$ ratio about 2.5 percent higher than that in the residues of organic material (see Figure 8-10). The $^{13}C/^{12}C$ ratio in the $CaCO_3$ from limestones of all ages is remarkably uniform; so also is the $^{13}C/^{12}C$ ratio in the kerogen found in shales. Departing from this pattern are the calcites associated with elemental sulfur. They have $^{13}C/^{12}C$ ratios similar to those in plant matter. We know from

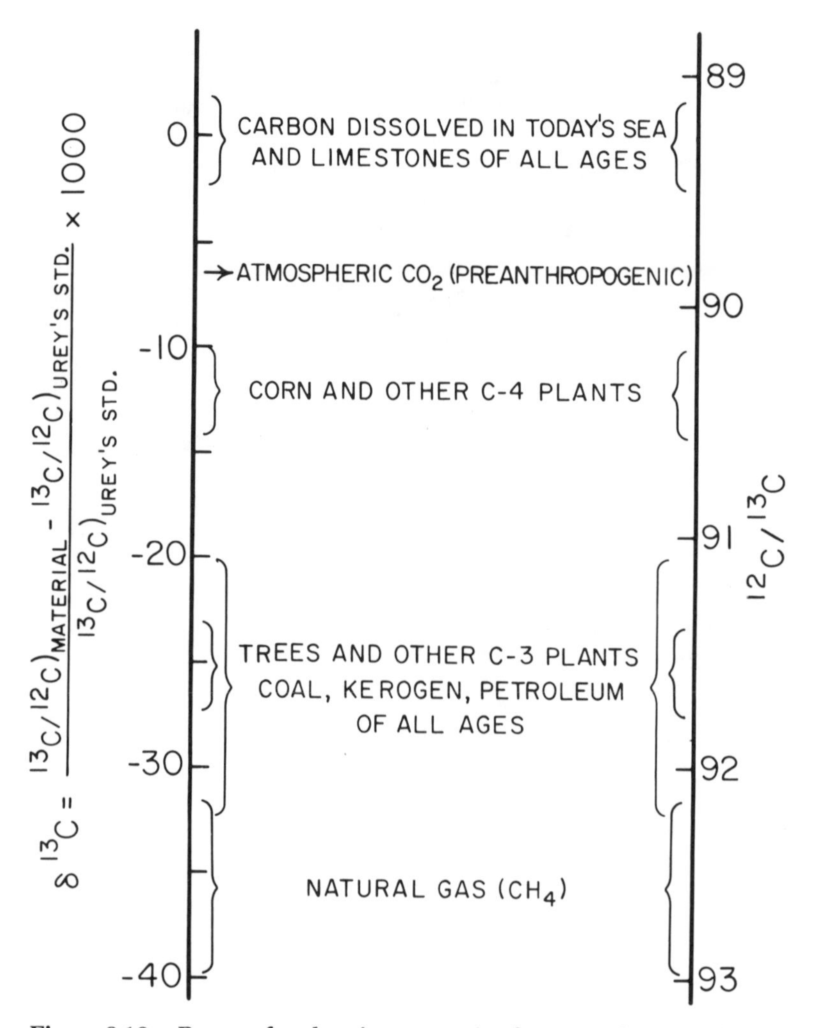

Figure 8-10. Range of carbon isotope ratios for natural materials: The scale on right shows the actual ratio of the light (^{12}C) to heavy (^{13}C) carbon atoms. The scale on the left was invented by stable-isotope pioneer and Nobel laureate Harold Urey. Geochemists find the parts-per-thousand difference scale handier for their purposes than the absolute ratios. Most plants living on the planet in the past, as well as today, use the Calvin photosynthetic mechanism (C-3). For this reason, coal and oil have carbon isotope ratios lying in the C-3 plant range.

experiments carried out in the laboratory that bacteria do not significantly separate the carbon isotopes during the conversion from methane to calcite form. Hence, the calcite appears to be telling us that its carbon atoms were derived from organic matter.

Fertilizer

In addition to CO_2 and H_2O, plants need nitrogen and phosphorus for growth. Both are drawn in via the root systems as dissolved constituents of soil water, the nitrogen as nitrate (NO_3^-) ion and the phosphorus as phosphate ($H_3PO_4^-$) ion. In natural ecosystems both of these elements are recycled over and over again (from living plant to dead plant debris to soil waters to living plant, and so on). The phosphorus in this recycling pool comes from minerals that undergo chemical weathering in the soil. The nitrogen comes from N_2 in the air. However, as mineral weathering and conversion of N_2 are processes that go slowly compared to the plant's need in many natural ecosystems, the availability of one or both of these ingredients limits plant growth. It is for this reason that nitrogen and phosphorus are the primary ingredients of fertilizer. In their attempt to produce more food per acre of land, farmers overcome the natural limits placed by the availability of nitrogen and phosphorus on plant growth by using fertilizer.

Some simple arithmetic shows that man has become dependent on fertilizer and that this dependency will become ever stronger. The land area of the earth is 15 billion hectares (the hectare is the metric unit of land area; one hectare is about 2.5 acres). About 10 percent (1.5 billion) of these hectares is now cultivated. With some difficulty another 0.8 billion hectares could be put into cultivation. The remaining 85 percent or so of the land surface is not arable; it is either too arid, too rainy, too rocky, or too cold. In the absence of fertilization an average of about 0.6 hectares of land is needed to produce enough food to sustain one human being. Hence, the presently cultivated land without fertilization would feed perhaps 2.5 billion people. Were all "arable" lands put to use the number would rise to perhaps 4 billion people. There are currently 5 billion people on the planet. Population will grow to at least 10 billion before it stabilizes. If starvation is to be fended off, the only answer is to increase the food yield from the average hectare of farmland. To do this requires massive use of fertilizer.

The N_2 contained in our atmosphere is an infinite resource in this regard. The problem lies not in the number of nitrogen atoms at man's disposal, but rather in the cost of converting N_2 to a form directly utilizable by plants. It lies in the strength of the chemical bond binding the N_2 molecule. Only a small fraction of the organisms living on our planet have the biochemical apparatus required to break the N_2 bond.This minority maintains the pool of fixed nitrogen (that is, the NO_3^- or NH_4+ forms utilizable by all plants). For man to break the N_2 bond in fertilizer factories requires the input of considerable energy (see Figure 8-11). Hence, food grown with fertilizer formed from atmospheric N_2 is more expensive than food grown on natural nitrogen. By the standards of industrial nations this extra cost is modest. By the standards of developing countries it is steep. Hence, while Earth has provided us with an atmosphere with more N_2 than we could ever use, we can only get at this one resource by using a sizable portion of another resource—fossil fuel.

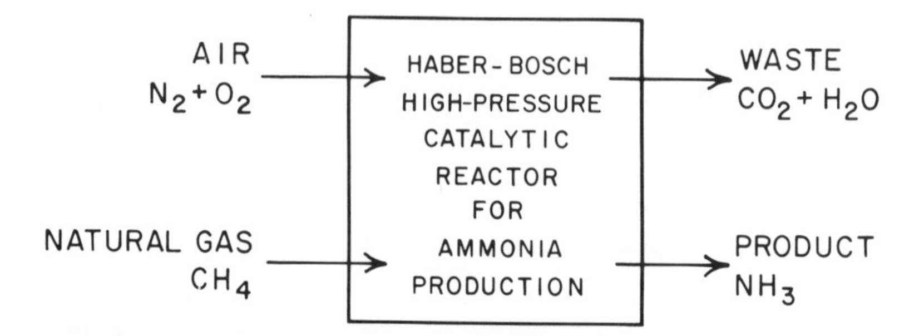

Figure 8-11. The basic process used by the fertilizer industry to "fix" atmospheric nitrogen: The ammonia produced in this way can be converted to a variety of salts (sodium nitrate, ammonium nitrate, urea, etc.) suitable for fertilizer. The methane consumed in this reaction constitutes the energy demand for this process.

One might ask if there are not reserves of nitrogen in other forms that could be converted to fertilizer at a lower cost. The obvious source is salts of nitrate and ammonia. These salts are very rare, however, because they are so soluble in water. The only important deposits are those formed from the droppings of bird populations that nest in very dry deserts and feed on fish from the adjacent sea. One such site is the Atacama Desert in Chile. Here birds feeding on the abundant fish population of the zone of upwelling off South America's west coast have formed mineable guano deposits. While of historic importance, these deposits currently meet only a very small fraction of the world's demand.

Earth was more generous with regard to phosphorus. Large sedimentary deposits of the mineral apatite have been located, the largest being in Florida. Man has only to bulldoze them up and do some relatively inexpensive grinding in order to ready them for addition to fertilizer. Like the other amazing concentrates, these deposits are a great boon to man. Average igneous rock and average shale contain only about 0.1 percent by weight phosphate. Were we forced to recover phosphate from such a source, it would be prohibitively expensive.

How did nature accomplish this feat of concentration? Surprisingly, we are not sure. Since the deposits we mine for phosphorus are geologically young, they must have formed under conditions similar to those of the present. Yet nowhere on Earth are similar deposits being formed today. This mystery has yet to be solved.

Summary

A common theme is shared by many mechanisms that generate nature's amazing chemical concentrates. The theme is *moving water*. Rivers carry the ingredients for evaporite deposits to the sea, and hot waters carry the ingredients for ore deposits. The waters move in one case because of energy delivered from the Sun and in the other case because of energy delivered by the decay of radionuclides. For both these mechanisms to operate, a planet must have water in liquid form pooled on its surface. While a few of these resources (diamonds, for example) might be found on a dry planet, most would be absent.

Supplementary Readings

Earth Resources, *by Brian J. Skinner, 1969, Prentice-Hall, Inc.*

An introductory treatment of metal, fertilizer, building material, energy, and water resources.

Hot Springs and the Ocean Floor, *by John Edmond and Karen Von Damm, 1983,* Scientific American, *V248.*

A summary of the chemical processes involved in the recirculation of seawater through the hot ridge crests.

Thermonuclear Explosion

Chapter Nine

Maintenance:

Mankind at the Helm

In the absence of "intelligent" life our planet would continue for billions of years along the same course it has followed in the past. Our Sun has enough hydrogen to burn smoothly for billions of years. The heat produced by the radioactivity of our Earth's mantle, although diminishing, will be adequate to drive the crustal plates for at least a few billion years.

Humans have changed the outlook for our planet, however. The first tool-bearing people appeared some 2 million years ago. Until about 10,000 years ago they competed on an equal basis with other Earth organisms and lived in harmony with the natural environment. Then they discovered agriculture, which gave them a great competitive advantage over their fellow animals. To enhance agriculture humans learned to reroute the flow of water for purposes of irrigation, to till the soil, to eliminate weeds and pests that threatened their agricultural lands. In so doing mankind initiated its tampering with natural cycles.

Starting about 150 years ago, industrialization greatly accelerated this man-made modification of our planet's surface. It has led to steady increases in the CO_2 content of the atmosphere, to the gradual acidification of our soils, and to the buildup of poisonous compounds throughout the environment.

Even more recently we mastered the secrets of the atom. Our nuclear arsenals now contain enough bombs to cripple our civilization and to threaten the very existence of many Earth species.

Thus, for better or worse, mankind is now at the helm. We have the power to modify climate, to determine the habitats of other species, to control the chemical purity of air and water. Unfortunately, our activities are intervening at such an alarming pace that we are not in control of developments. How much damage will be done before we are able to take the reins in a responsible manner?

Introduction

Is there any reason to believe that our planet's hospitality toward life will someday change for the worse? In the absence of mankind the answer to this question would likely be no. Our Sun will burn steadily for several billion years before becoming depleted of its nuclear fuel. Hence, climate will fluctuate around a reasonably congenial mean in the future much as it has in the past. While extremes of these changes frequently threaten vulnerable forms, the advancement of life has not been stifled by climate. The Earth's internal heat source—the decay of long-lived radionuclides—although slowly diminishing, will remain strong enough to drive the movement of the mantle's convection cells for a billion or so years. Hence, the beauty of the landscape, the energy and chemical resources, the soils, and the constituents of the atmosphere and ocean will continue to be renewed. Of course, another asteroid or comet might bring a repeat of the disastrous impact that occurred 65 million years ago. Were this to happen many organisms that now inhabit the Earth would become extinct. Those that survived, however, would eventually repopulate the planet. In spite of missing a beat, the evolution of life would march on.

These generalizations do not apply when human beings are considered. During the last hundred years we have significantly modified our planet's environment. Many of our major rivers are now impounded behind great dams. Ten percent of the land now supports man's agriculture rather than the natural flora. Deserts are expanding as adjacent lands are overgrazed. A number of forms of life have become extinct. Many others are threatened. Our atmosphere now contains traces of freons and other man-made gases. Our soils erode at an accelerated pace. Our fossil-fuel resources will be largely consumed within two more centuries. We have the explosive force in our nuclear arsenals to destroy all the Earth's major cities and industrial facilities. Were a full nuclear exchange to occur, many believe the skies would be temporarily blackened by smoke from the resulting fires. Without light, plants could not grow. Without plant food, animals would starve. It would be cold. In addition, the Earth would be blanketed by radioactive fission products whose decay would induce immense genetic damage. At least for the immediate future our planet's habitability lies in our hands!

Climate Modification from Greenhouse Gases

A number of mankind's activities could lead to significant changes in climate. The largest of these impacts will probably come from the extra CO_2 we are adding to the atmosphere, generated both by the burning of fossil fuels and the destruction of forests. While in the atmosphere, CO_2 is chemically inert. Hence, the CO_2 we release accumulates. From studies of the air trapped in bubbles in polar ice cores it is estimated that the CO_2 content of the atmosphere was about 280 parts per million (ppm) and nearly constant during the centuries prior to 1800. In 1958, when the first sufficiently accurate direct measurements were made, the atmosphere's CO_2 content had already climbed to 316 ppm. As of 1985 it was about 345 ppm (see Figure 9-1). By the end of the next century the atmosphere's CO_2 could well be more than double its preindustrial value! A possible scenario for this continued buildup is shown in Figure 9-2. This increase in CO_2 is too small to have an impact on the atmosphere's bulk. In fact, for each new CO_2 molecule that appears, one O_2 molecule disappears (the 65 ppm rise in CO_2 content between 1800 and today has been balanced by a comparable decrease in O_2). The concern comes from the fact that while O_2 molecules are transparent to the passage of outgoing earthlight, CO_2 molecules are not. Thereby, as discussed in Chapter Seven, they impede the return to space of the energy received from the Sun. The Earth's surface promptly responds to the imbalance in its energy budget created in this way by becoming a bit warmer. A warmer Earth emits more infrared light. In this way the retardation in the loss to space of earthlight created by the excess CO_2 molecules is compensated. Like the glass in a greenhouse, CO_2 creates a warmer environment. Because of this similarity, CO_2 buildup is often referred to as the "greenhouse effect." The current estimate is that when the atmosphere's CO_2 content reaches 600 ppm (about twice the preanthropogenic

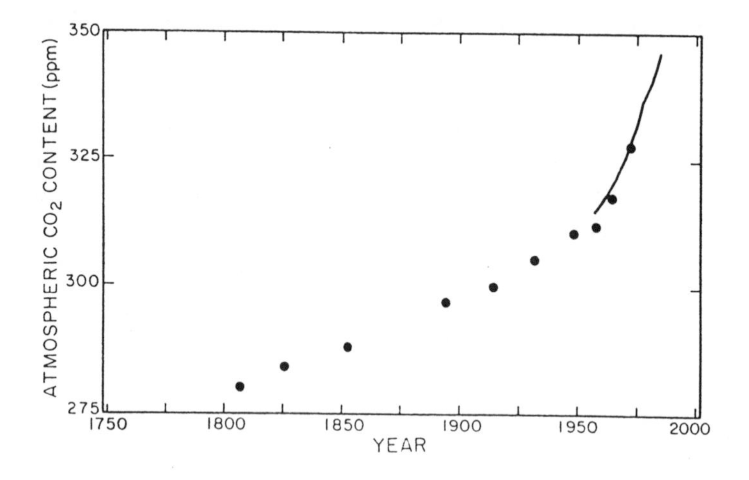

Figure 9-1. Variation in the atmospheric CO_2 content over the past 200 years: The dark line represents the results of a continuous measurement series extending from 1958 to the present made by Charles Keeling of Scripps Institution for Oceanography and his co-workers. The circles represent measurements made by Oeschger and his colleagues at the University of Bern on air bubbles trapped in ice from Antarctica.

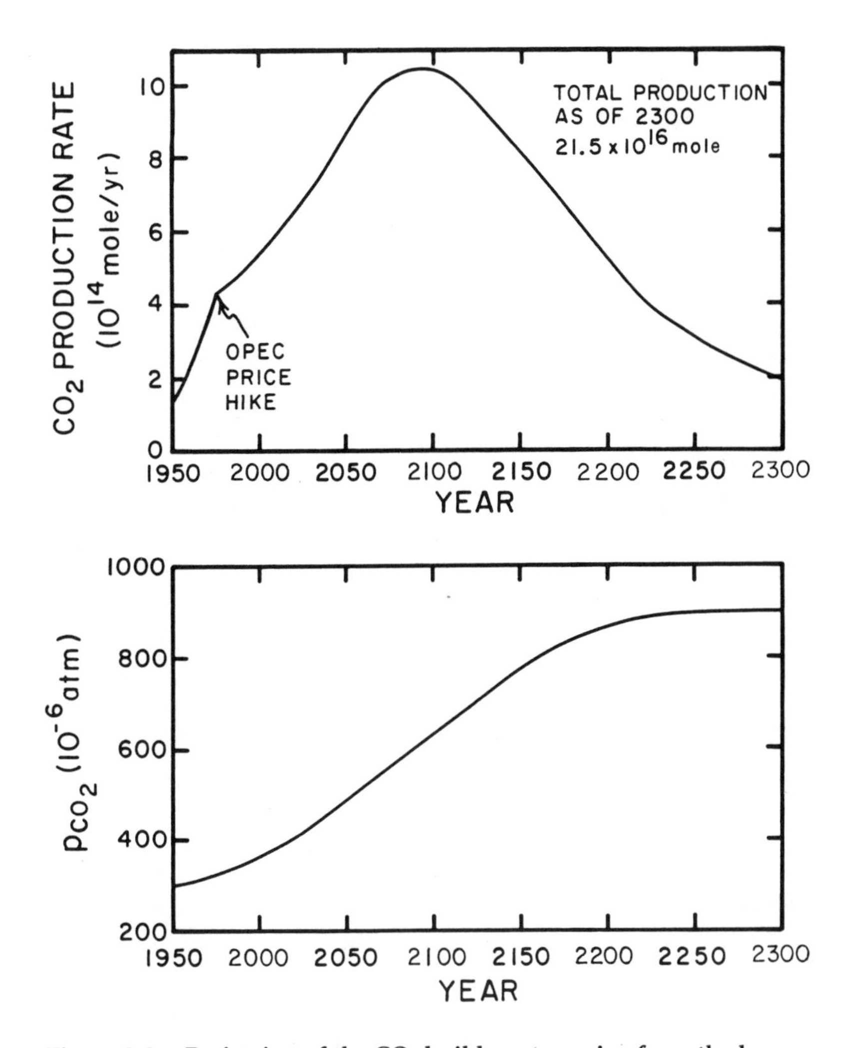

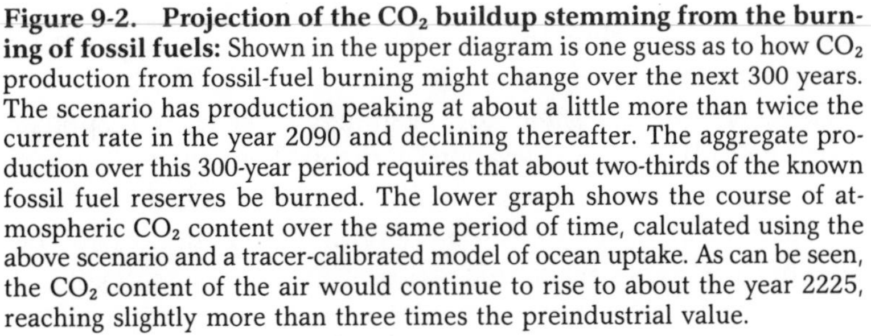

Figure 9-2. Projection of the CO_2 buildup stemming from the burning of fossil fuels: Shown in the upper diagram is one guess as to how CO_2 production from fossil-fuel burning might change over the next 300 years. The scenario has production peaking at about a little more than twice the current rate in the year 2090 and declining thereafter. The aggregate production over this 300-year period requires that about two-thirds of the known fossil fuel reserves be burned. The lower graph shows the course of atmospheric CO_2 content over the same period of time, calculated using the above scenario and a tracer-calibrated model of ocean uptake. As can be seen, the CO_2 content of the air would continue to rise to about the year 2225, reaching slightly more than three times the preindustrial value.

value), the mean global temperature will be 3° to 4° centigrade higher than now. A warming almost as great as that which accompanied the transition from the last period of glaciation to the present period of interglaciation (see Figure 9-3).

Since we have no recent geologic analogue of global climates significantly warmer than today's, there is no ready way to assess the environmental impact of such a warming. Our insight must come mainly from climate simulations designed to duplicate the current meteorological conditions. It is from these models that the estimate of a 3° C to 4° C warming for doubled CO_2 is obtained. While the models produce myriad regional details, most of these are not taken as serious predictions of things to come. The reason is that, despite the best efforts of the modelers, the simulations are very likely flawed. For example, the models employ highly simplified means for generating clouds, because our knowledge of the factors controlling cloud distributions, heights, and densities is rather poor. So also is our knowledge of the physics of the processes that create the water droplets and ice crystals of which clouds are made. Cloudiness is very important to climate, because clouds reflect away substantial amounts of sunlight and thereby cool the Earth's surface. Hence, changes in cloudiness brought about by a stronger greenhouse effect must be properly modeled.

The models also carry only very primitive simulations of the interactions between the atmosphere and the oceans. As the oceans currently carry a great deal of heat from the tropics to higher latitudes, the response of the ocean to changing atmospheric CO_2 content is an important part of the problem. Again, we do not know enough about the ocean to properly model it. Because of this and other basic deficiencies, predictions of regional climates provided by these models must be taken with a grain of salt.

These uncertainties lead to a hesitancy in society's response. Too often misled by exaggerations of the likely impacts of various pollutants, politicians are not willing to give the CO_2 question serious consideration. While supplying token support for research, they do not place this problem in the same category as

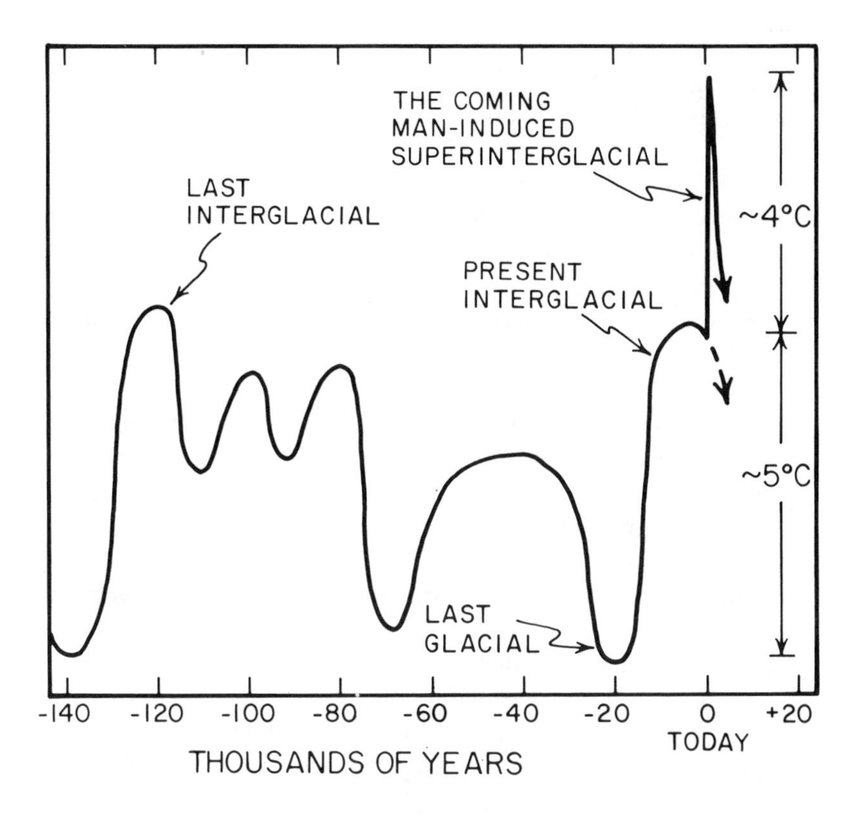

Figure 9-3. The coming superinterglacial: The projected CO_2 warming is compared to what we know about the Earth's temperature history for the last 140,000 years. The warming will likely be as great as that which accompanied the demise of the last great ice sheet (12,000 years ago). It will be far more abrupt, however! The rise will occur in a few centuries. Then, once we have run through our fossil-fuel reserves, the Earth's temperature will slowly come back down as the excess CO_2 is taken up by the ocean and eventually deposited as $CaCO_3$ on the sea floor. Were man not to have intervened with his greenhouse gases, the expectation is that the present interglacial would have come to an end in a fashion similar to that seen following the last interglacial (dashed arrow).

defense against nuclear attack or prevention of death by cancer. Our leaders seem to be willing to gamble that the impacts of CO_2 will be sufficiently small that we can cope with them in much the same way that we now cope with droughts and other unfavorable weather perturbations. The threats posed by nuclear weapons and by cancer are clear to everyone. By contrast, even on a long time scale, the buildup of CO_2 does not pose any concrete threat that even scientists can agree upon. They can say only that there is a significant likelihood that major environmental changes will take place and that the impact of these changes will vary with geographic location. To some the changes will be a blessing, to others they will be a curse, and to still others they may go largely unnoticed.

Can we stem CO_2 production? This would be the ideal solution. Such an action, however, is not feasible. Mankind is now totally dependent on energy, and most of our energy for the next century at least will have to come from the burning of fossil fuels. Furthermore, there is no practical way to capture the CO_2 produced when these fuels are burned. Unlike SO_2 and N_2O, which are minor constituents of exhausts, CO_2 is a major constituent. For each 100 kilograms of gasoline burned in your automobile, about 300 kilograms of CO_2 is produced! Nor can we reclaim CO_2 from the atmosphere. The amount of energy required to recover a kilogram of CO_2 would be comparable to the amount of energy gained by producing this CO_2 in the first place. Hence, to the extent to which we are dependent on energy, we are destined to have the CO_2 content of atmosphere rise.

Uptake of Fossil Fuel CO_2 by the Ocean

Between 1958 and 1984 the rise in CO_2 content of the atmosphere was only about half that expected if all the CO_2 generated by the combustion of fossil fuels were to have remained in the atmosphere. From this observation it is clear that some of the excess CO_2 is going elsewhere. The most obvious candidate for elsewhere is the ocean. Were the CO_2 we have produced to be thoroughly equilibrated between ocean and atmosphere, then only 1/6th would remain in the atmosphere and 5/6ths would have taken up residence in the sea. It should be emphasized that CO_2 is unusual in this regard. For other man-made gases, such as radiokrypton (a by-product of nuclear reactors) and the freons (widely used as refrigerants), the split would be 19/20ths in the atmosphere and only 1/20th in the sea. CO_2

differs because it undergoes a chemical reaction with the carbonate ions ($CO_3^{=}$) dissolved in the sea to form bicarbonate ions (HCO_3^-). The reaction is as follows:

$$CO_2 + CO_3^{=} + H_2O \rightarrow 2HCO_3^-$$

The ocean is naturally endowed with a sizable component of $CO_3^{=}$ ion which enables it to take up an appreciable amount of the CO_2 we produce (see Table 9-1). Were the ocean nearly free of $CO_3^{=}$, as are most of our lakes and rivers, then the equilibrium distribution for excess CO_2 between atmosphere and ocean would be more similar to that for normal gases.

If the sea is capable of taking up five-sixths of the fossil-fuel CO_2 we have produced, then why does one-half remain in the air? The reason is that the sea does not mix very rapidly. Our knowledge of the rates of vertical stirring in the sea comes largely through the distribution of radioisotopes. Oceanographers quickly adopted Willard Libby's radiocarbon dating method to

Table 9-1. Summary of the sizes of Earth's carbon reservoirs and the changes in these sizes caused by man's activities:

Material	Amount in 1800*	Amount in 1985*	Man-made change*
Limestone + dolomite	240,000	240,000	—
Kerogen	60,000	60,000	—
Coal + lignite†	31	30	−1.1
Oil†	2.3	1.9	−0.4
Natural gas†	1.3	1.2	−0.1
Ocean HCO_3^-	270	270	+0.6 (Ocean HCO_3^-, $CO_3^{=}$, CO_2 combined)
Ocean $CO_3^{=}$	30	30	
Ocean CO_2	10	10	
Atmos. CO_2	4.5	5.5	+1.0
Soil humics	2	2	?
Living biosphere	0.5	0.5	?

*The amounts are in units of 10^{20} moles.

†Known reserves.

assess the renewal rate of waters in the deep sea. Radiocarbon atoms are continually produced by nuclear reactions triggered by the collisions of galactic cosmic rays with our atmosphere. After a mean lifetime of 8200 years these radiocarbon atoms decay to the stable nitrogen form from which they originated. Because their mean life is an order of magnitude longer than the time required for the ocean to mix, radiocarbon atoms are found throughout the sea (as CO_2, HCO_3^-, and $CO_3^=$). Their distribution is not uniform, however. The carbon dissolved in surface waters has a higher ^{14}C content than the carbon found in deep waters. Oceanographers have mapped these differences and use them to provide estimates of the rate of renewal of deep waters in various parts of the sea. As can be seen in Figure 9-4, these times range from a few hundred years for the deep northern Atlantic to almost 2000 years for the deep northern Pacific.

Even more useful for estimating the extent to which the ocean has equilibrated with the excess CO_2 in the atmosphere is the distribution of man-made tracers in the sea. The key tracers are tritium (produced by nuclear tests carried out in the atmosphere)* and freons (used as refrigerants, propellants, and foaming agents). It is as if a red dye were spread across the entire ocean surface and monitored as it spread downward into the ocean's interior. By measuring the vertical profiles of tritium, oceanographers have been able to determine the rate at which substances penetrate into various regions of the sea (see Figure 9-5). They find that one decade after the tritium was added it had spread through about one-tenth of the ocean's volume. Based on these results it is possible to assess the extent to which fossil-fuel CO_2 has penetrated into the sea.† The result obtained

---

*Tritium is an isotope of the element hydrogen. It has two neutrons and one proton in its nucleus. It is radioactive with a half-life of 12.2 years.

†One might ask why the distribution of fossil-fuel CO_2 molecules is not directly measured in the sea. The problem is that even in the surface ocean the excess CO_2 constitutes only about 2 percent of the total dissolved carbon. Unfortunately none of the pre-1970 measurements of carbon in seawater was made to better than $\pm$ 1 percent accuracy. Thus, there is no direct way to measure the excess.

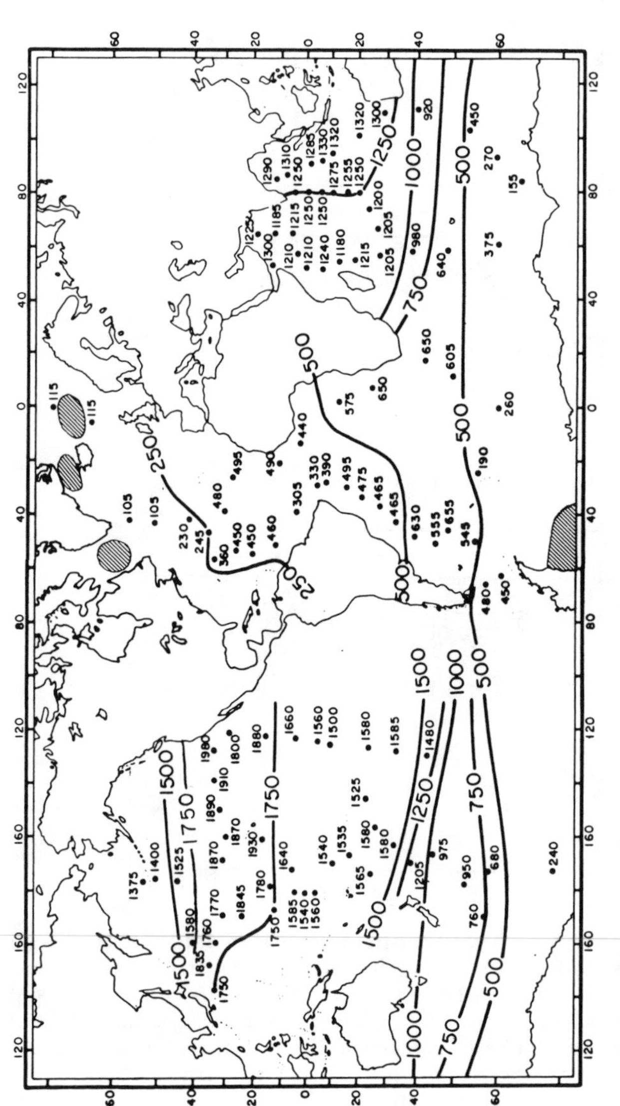

Figure 9-4. Radiocarbon ages for deep-sea water: Shown on this map are the stations occupied during a large survey of the ocean carried out in the 1970s. The $^{14}C/C$ ratio was measured in about 20 samples from throughout the water column at each of these stations. The apparent age of the water at 3 kilometers depth calculated from these radiocarbon results is shown. Since new deep water forms in the northern Atlantic (see shaded areas) but not in the northern Pacific or Indian oceans, the water in the Atlantic Ocean is much younger than the water in the deep Pacific and Indian oceans. The average radiocarbon age for deep-sea water is about 1000 years.

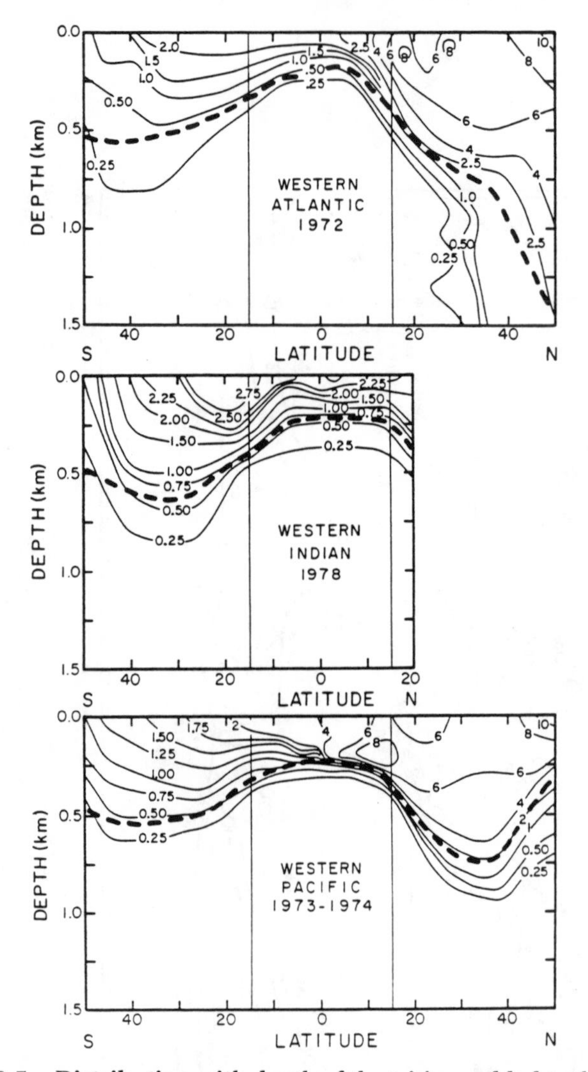

Figure 9-5. Distribution with depth of the tritium added to the ocean as the result of hydrogen bomb testing: This tritium reached the sea surface mainly in the period 1962 to 1965. The surveys were carried out about one decade later. As can be seen, more tritium reached the northern hemisphere than the southern hemisphere. This is because the tests producing the bulk of the tritium were conducted in the northern hemisphere. The heavy dashed line shows the depth at which the tritium concentration reaches one-quarter of the value in overlying surface water. Since the mean depth of the ocean is 3800 kilometers, this survey tells us that on the time scale of a decade substances added to the sea surface are stirred into only about one-tenth of the sea's volume.

is that 40 percent should be in the sea and 60 percent in the atmosphere. This result is consistent with the observation that 55 percent of the CO_2 produced since 1958 remains in the air.

Other Greenhouse Gases

Of Earth light-absorbing gases, CO_2 is not the only one whose atmospheric content is rising. Another is methane (CH_4). Because methane is not stable in the presence of oxygen gas, the average molecule survives only about ten years in the atmosphere before it reacts with O_2 to form CO_2 and H_2O. Because of this, the concentration of methane in the atmosphere is quite small (only 1.6 parts per billion). However, methane is an effective Earth light-absorber. Thus despite its small concentration it is a significant greenhouse gas.

Measurements on polar ice show that the methane content of the atmosphere is on the rise. It has doubled since the year 1800 (see Figure 9-6). This increase is thought to reflect both an increase in the production rate of methane and an increase in its atmospheric lifetime. The increased production is attributed partly to the steep rise in cattle population (what might be called the McDonald's Syndrome); like fellow ruminants, cattle convert part of their food carbon into methane rather than carbon dioxide. Another increasing source is rice paddies. Methane is produced in the O_2-free mud in which rice grows. From here it seeps up through the rice stalks into the air. The increased atmospheric lifetime of methane molecules is attributed to a decrease in the ozone content of the Earth's lower atmosphere. As ozone is the catalyst for methane destruction, lower ozone contents lead to lower destruction rates.

The atmospheric content of freon and of other industrial gases is also on the rise. Since most of these gases are not naturally produced, we don't need polar ice measurements to establish their preindustrial levels. It was zero. While eventually destroyed, gases like freon survive in the atmosphere for times on the order of 100 years.

We know too little about the expected production rates of these gases to adequately chart their future greenhouse impact.

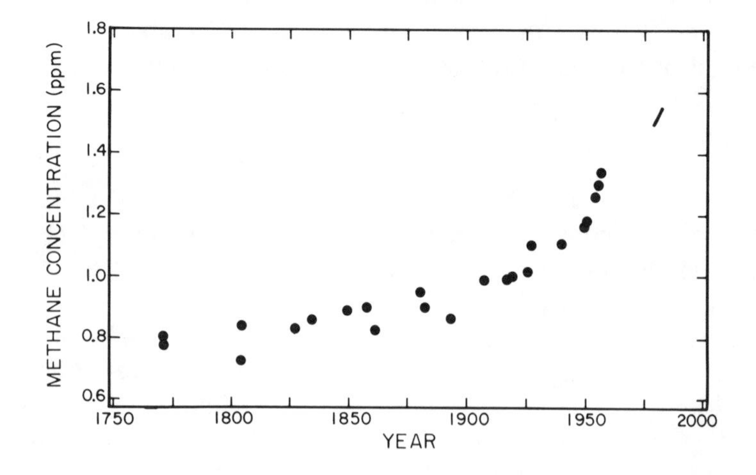

Figure 9-6. Methane content in air over the last 200 years: The circles represent measurements made on ice cores. The short line segment shows the results of direct measurements on air samples. As can be seen, the methane content of the atmosphere has doubled since the beginning of the industrial revolution.

However, the estimates available suggest that their combined greenhouse power will be about one-half that for excess CO_2.

Coping with the Greenhouse Warming

It seems inevitable that the amount of energy we use and the source of this energy will be dictated by economics. As long as the number of people on the planet continues to increase, so will our energy-use increase. So also will the demand for beef and rice. So what should we do? Since we can't significantly influence the amount of these gases that are produced or their fate once released, our strategy must be rather to prepare to cope with the changes the greenhouse effect is likely to bring. The key to coping is foreknowledge. With some warning we can create strategies designed to soften the curses and to enhance the blessings that increased greenhouse power will bring.

The situation is much akin to that faced by an engineer hired to manage a complex chemical plant. During his first few months on the job the engineer gets the impression that he has inherited an operation requiring little of his attention. However, as time passes he notes some odd fluctuations in the plant's usually smooth flow. Also, now and then he has to carry out small repairs. Feeling a bit less secure, he decides that he should have a look at the plant's blueprints and operating manuals. To his amazement none can be found. His inclination is to reconstruct these plans based on what he sees. When he attempts to do this, however, he finds that he is in for a long and difficult task. Instead he decides to shelve this project in the hope that he can keep things running through stopgap measures. As the years pass, the plant is stressed by demands for higher output and by complaints regarding the adverse effects it is having on the environment. The beleaguered manager does his best to cope with these conflicting forces, but he realizes that the system is being pushed toward operating conditions quite different from those for which it was designed. He struggles to make changes designed to fend off breakdowns and reduce the complaints from his neighbors. The changes prove very expensive and often fail to accomplish their objective. Looking ahead he sees the situation worsening. Finally he bites the bullet and initiates a major effort to reconstruct the operational plans for his plant. Having grown in wisdom, he realizes the importance of distinguishing between the "normal" fluctuations experienced by the plant and those related to the new stresses placed on it by the demands for higher output. His hope is that this effort will allow him to make some valid predictions regarding the plant's future. The predictions, he reasons, would permit cost-effective planning for the plant's maintenance. He would also be better able to judge the likelihood that conditions would arise which would threaten the plant's continuity of operation. He launches this project with a sense of urgency.

We stand at a similar moment of decision. We have only begun the very difficult task of understanding how the Earth operates as a joint physical, chemical, and biological system. It will be a very long, very difficult, and very expensive job. Other priorities seem more pressing. Nonetheless, cost-effective management of our planet makes it essential that we take the plunge.

Purposeful Climate Modification

By modifying the Earth's greenhouse power, man eventually will learn the sensitivity of the environment to changes in the radiation budget. With this knowledge will come a great temptation. Proposals will be put forth of ways to engineer an even more equitable climate than we now enjoy or to ameliorate the adverse effects of the increased greenhouse power.

So far little has been written on this subject. It is one of the few subjects considered taboo in the realm of scientific inquiry (ranking perhaps number two after inquiries into possible racially related differences in intelligence). This unofficial ban is likely to disappear as soon as the buildup of greenhouse gases produces significant climate changes; the governments of adversely affected countries will demand action.

Are such modifications feasible? Budyko, a Russian meteorologist, was the first to propose a possible scheme. He pointed out the possibility that greenhouse warming might be countered by aerosol cooling. Budyko suggested that by adding SO_2 gas to the upper atmosphere (stratosphere), sunlight-reflecting aerosols could be generated. With the right amount of aerosols, the greenhouse warming could be countered.

American scientists have attempted to quantify Budyko's proposal. They estimate that 35 million tons of SO_2 gas would have to be transported to the stratosphere each year in order to counterbalance the warming produced by doubled atmospheric CO_2 content. The addition would be repeated each year, because the aerosols produced by the SO_2 would remain in the atmosphere for only about one year. At current prices, the cost of this SO_2 would be 15 billion dollars per year. If carried to the stratosphere in Boeing 747 airplanes, the delivery cost would be 15 billion dollars. The military defense budget for the U.S. is now about 350 billion dollars per year. By these standards 30 billion dollars per year for climate defense is not beyond our means. Hence, Budyko was right. At least one economically feasible strategy for purposeful global climate modification exists. Other strategies will probably be found.

How does the Budyko scheme work? SO_2 is unstable in the atmosphere. It reacts (via the ozone catalyst) to produce sulfuric acid (the noxious component of acid rain).* Sulfuric acid is not a gas. Rather, it agglomerates into aerosols (very small droplets). These aerosols have curious optical properties: they reflect sunlight but not earthlight. Hence, the H_2SO_4 aerosols reflect away incoming sunlight without interfering with outgoing earthlight. One might liken them to a one-way mirror.

We know about this phenomenon because nature frequently performs experiments. The last one was in 1982 when the Mexican volcano El Chichon erupted explosively and propelled about 8 million tons of SO_2 gas into the stratosphere. Through a combination of satellite- and Earth-surface-based observations it was possible to assess how much SO_2 was delivered, how fast this SO_2 was converted to H_2SO_4 aerosols, and how long the H_2SO_4 aerosols remained in the stratosphere. It was also possible to determine the size spectrum and optical properties of these acid aerosols.

The point is not that Budyko's strategy is necessarily a wise one, but, rather, that purposeful global climate modification lies within our grasp. One hundred years from now the temptation to take such an action may be high. How we respond to this temptation is certain to influence the future of our planet.

Energy Production Over the Next Century

Our reserves of oil are adequate for only about a hundred years. After that our automobiles, airplanes, and heating plants will have to be fueled in some other way. While the situation for natural gas may be a bit better, its supply will surely run short by the end of the next century. Once these energy sources run low, we will have to turn to less desirable fuel sources. Coal is number one on the list and ^{235}U is number two. Together these two sources could fuel the world for a few centuries.

*Industry now adds 350 million tons of SO_2 each year to the lower atmosphere from its smokestacks. As this SO_2 and the sulfuric aerosols it creates remain airborne for only a few days, its impact on climate is small.

From an environmental point of view, neither coal nor uranium is highly desirable. To recover coal either we have to risk the health and lives of the miners or we have to strip away overlying sediments to get at the coal without mining. The stripping procedure, no matter how carefully conducted, mars the landscape. In addition to the drawbacks associated with the recovery of coal, there are drawbacks associated with burning it. We have already discussed CO_2. Since about 80 percent of our fossil-fuel reserves are as coal, this option will have a big greenhouse impact. In addition to CO_2, coal-burning produces SO_2 which rains back to the surface of the Earth as sulfuric acid. The acid eats into the stone monuments of our civilization, it makes many of our lakes uninhabitable for fish, it appears to be taking a toll on our forests, and it may eventually damage our soils. In addition, coal-burning releases selenium, cadmium, arsenic, and other toxic metals. While these poisonous byproducts of coal-burning can be captured in the smokestacks, the capture process is expensive. Hence, coal-burning will add to the planet's growing burden of environmental poisons. Oil and especially natural gas contain far lower levels of sulfur and arsenic and, hence, by comparison to coal, these fuels are environmentally clean. Also, the recovery of oil and gas is safer and has far fewer environmental impacts. So if we are forced to choose coal it will definitely be a step backward in our campaign to maintain a "clean" environment.

Roughly 15 percent of our electricity is now obtained from nuclear powerplants. In these plants ^{235}U is fissioned. The energy released by fission is converted to electricity. Three major threats are posed by this activity. First, for each ^{235}U atom fissioned, two radioactive atoms are produced. Some of these fission products have half-lives sufficiently long that the storage of nuclear ashes has become a major issue. Were society's energy demands maintained by nuclear power for a century, the disposal of these radioactive wastes would pose an enormous problem. Despite three decades of research, no disposal procedure satisfying to all interested parties has been found. Either they are too expensive or too risky!

The second danger posed by nuclear power has to do with reactor malfunctions or sabotage which could lead to a so-called meltdown followed by an explosion. The radioactive debris from such an explosion could render an area many tens of miles in diameter uninhabitable for many decades. The movie *China Syndrome* and the real-life Three Mile Island incident made it clear to the public that such accidents must be taken more seriously than utility companies have led us to believe. As this book was being completed the world was shocked by the meltdown of the nuclear reactor at Chernobyl in the USSR. While the impact of this disaster had yet to be fully assessed, it provided a dramatic demonstration of the risks associated with nuclear power generation.

The third concern has to do with the plutonium that is produced as a by-product of the operation of all reactors. Unlike uranium itself, plutonium can be converted rather easily to nuclear weapons. The fear is that the more nuclear facilities there are in operation, the greater the likelihood that more governments and even terrorist organizations will produce atomic bombs.

Clearly, both uranium and coal will be used. On the short term we have no other viable option. The ratio of coal use to uranium use in any given country will depend on local economics and political attitudes. At the time this book was written the resistance to nuclear power development ran high. People were more willing to accept the long-experienced hazards of coal than the newer, and seemingly more acute, hazards of uranium. Fortunately, neither threatens the existence of life. However, the chemical "ashes" of coal power and the nuclear "ashes" of uranium power are bound to leave some black marks on our planet.

The greatest energy-related danger during this period is, of course, that posed by the nuclear arsenals of the superpowers. A full exchange of these missile-delivered hydrogen bombs would have awful and far-reaching consequences to all the inhabitants of the planet.

Longer-Term Energy Options

Together, conventional coal and uranium plants cannot fuel the world for more than several centuries. During that time a new energy source will have to be put into large-scale operation. We appear to have three options. The first and most ideal would be to directly harness the energy delivered by the Sun. This resource is unlimited. The problem is that it is dilute. Even with the most optimistic conversion system, sizable areas of the Earth's surface would have to be covered with solar panels. While such panels are now made to power space probes and communication satellites, they are extremely expensive and have limited lifetimes. Solar electricity is currently 100 times as expensive as electricity from fossil-fuel-powered plants. Major technologic breakthroughs will have to occur before this approach becomes economically feasible.

A second option is fusion power. As does our Sun, powerplants could conceivably derive energy from the fusion of light elements into heavier ones. The principles are understood. The needed reactions have been carried out in the laboratory. The problem lies in the required temperature. As discussed in Chapter One, the fusion of hydrogen atoms in stars requires temperatures in excess of 50 million degrees. No material known to man can withstand temperatures above even several thousand degrees. So what kind of bottle can be created to hold the fusion fire? One possible solution is to use a magnetic bottle in which the hot gas is prevented from touching the walls. The hot gas is held in a cold vessel. Such bottles have been made but they are enormously expensive and very inefficient. The energy required to run them still exceeds the energy released by the nuclear reactions occurring within them. In other words, the prototypes are energy sinks rather than energy sources. While the expectation is that improvements can be made that will reverse this situation, there still is no way to know when an operable large-scale system can be built and how much the energy produced in this way will cost. Time will tell.

The third route is a more certain one but a more dangerous one. It involves what is called the breeder reactor. Conventional power reactors use only ^{235}U. For every single ^{235}U atom on the Earth, there are 139 ^{238}U atoms and about 400 ^{232}Th atoms. In conventional nuclear reactors these other two isotopes go unused because they do not fission when struck by a neutron (as does ^{235}U).

In the breeder reactor ^{238}U or ^{232}Th are converted to fissionable nuclides by bombardment with neutrons released during the fission of ^{235}U atoms. On the average, three neutrons are released when a ^{235}U atom fissions. Only one is needed to sustain the chain reaction. The remainder are largely sopped up in the reactor's control rods (a small percent by chance interact with the ^{238}U present in the uranium fuel, leading to the production of plutonium). The breeder reactor is designed to maximize the production of fissionable material. The idea is that it produces one or more fissionable atoms from ^{238}U and ^{232}Th for each ^{235}U atom consumed (see Figure 9-7). The fissionable material made in the breeder can then be used as fuel. It too will breed its replacement. Theoretically, by this scheme, we could utilize all the ^{232}Th and ^{238}U as well as the ^{235}U atoms. By this method the world's energy needs could be met for an indefinite period.

The large-scale use of breeder reactors would greatly aggravate all of the concerns expressed above for the conventional reactor. More nuclear ashes would be produced. Because the breeder is a more complex reactor, the chances for malfunction go up. Most serious, the major fuel would become plutonium, and the large amounts of this substance being handled each year would raise the likelihood that some of it would be diverted and used for clandestine purposes.

Soils

Of all man's resources, the least appreciated is soil. Much of Earth's land surface is mantled by a layer of chemically altered rock fragments mixed with the organic residues of plant growth. This layer holds the water and nutrient supply for plants. Besides light, water, nutrients, and equitable temperatures, plants require proper soil to house their roots.

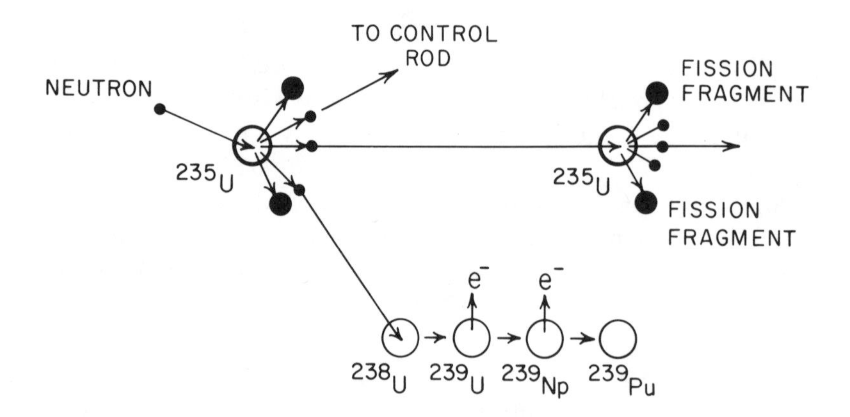

Figure 9-7. A scheme for breeding nuclear fuel: The goal is to put to use the extra neutrons released during fission. Since only one is needed to sustain the reaction, the other two can either be wasted (by capture in the material in the reactor's control rods) or be used to upgrade the more abundant nonfissionable ^{238}U isotope to a fissionable form. In this way the energy potential of uranium could be increased a hundredfold! In the scheme shown above, for each ^{235}U atom which fissions one neutron goes to maintain the chain reaction, one is absorbed by a ^{238}U atom (making it into a ^{239}U atom), and one is absorbed in a control rod. The ^{239}U atom undergoes radioactive decay, becoming ^{239}Np. The ^{239}Np atom undergoes radioactive decay, becoming ^{239}Pu. This plutonium atom is fissionable. Thus, by this process, for each ^{235}U atom consumed one nonfissionable ^{238}U atom is converted to a fissionable plutonium atom.

Man's takeover of the Earth's lands has not been particularly good for soils. Our practices have led to greatly accelerated erosion. As much as 5 percent of the soil present on the planet when man commenced agriculture, grazing, forestry, and irrigation has been carried away by wind and water. The rate of erosion on lands manipulated by man has run ten to a hundred times the natural erosion rate. At this rate we could lose a sizable portion of our soils over the next several hundred years.

In addition to being lost to erosion, our planet's soils are undergoing man-induced chemical alterations. Irrigation often leads to salt buildup in soils. Acid rain mobilizes aluminum and other metals in the soil.

One might conclude that if we lose our soils to erosion or contaminate them chemically, we can always make new ones. Dirt is cheap. Even if artificial soil were cheap (and it isn't), a staggering total of about 3000 metric tons of soil is required per hectare of land. So if you had to manufacture and transport soil to make a plot of land capable of growing your own food it would be a lifetime of backbreaking labor. Thus, a key part of any long-term management plan for the planet must be to protect our soil resource.

Summary

Suddenly, without much warning, man finds that his activities have side effects that are not trivial compared to the natural processes that maintain the environment we enjoy. These side effects may significantly alter our climate and our soils. We also find that we are running rapidly through the cream of nature's chemical concentrates. Once this cream has been skimmed, we will have to turn to less desirable resources to fill the energy and ore gaps that will arise. Instead of attending to the proper management of our planet's future, we are obsessed with the nuclear arms race, with extending our lives through medical technology, and with entertaining ourselves with gadgets. Little has changed since Roman times. Man fiddles and hopes that somehow the future will take care of itself. It surely will, but mankind may not like the course it takes. Let us hope that things will soon change, and that humans will accept their responsibility to maintain planet Earth.

Supplementary Readings

Energy and Power, *1971, A Scientific American Book, W. H. Freeman and Company.*

A series of essays on energy sources and energy use.

Food and Agriculture, *1976, Scientific American.*

A series of essays on agricultural systems.

Energy and Environment, *edited by Raymond Siever, 1974, Readings from Scientific American, W.H. Freeman and Company.*

A series of essays on energy production systems and their environmental impacts.

Natural Resources for U.S. Growth: A Look Ahead to the Year 2000, *by Hans H. Landsberg, 1964, Johns Hopkins University Press.*

A comprehensive treatment of the availability of resources in the United States.

Tracers in the Sea, *by Wallace S. Broecker, 1982, Eldigio Press.*

A treatment of chemical processes taking place in the ocean, including the uptake of fossil fuel CO_2.

Glossary

Ultrasmall units of matter

Atom—the basic chemical unit of matter; consists of a dense core of neutrons and protons orbited by a cloud of electrons.

Electron—one of the basic units of the atom; characterized by a very small mass (compared to the proton and neutron) and by a negative electrical charge.

Element—the chemist's subdivision of atoms; based on number of protons in the atom's nucleus.

Ion—an atom in which the negative charge carried by the orbiting electrons does not match the positive charge of the protons in its nucleus.

Isobar—one of the subdivisions of nuclides used by the physicist; nuclides with the same total number of nuclear particles but different proton and neutron numbers are isobaric brothers.

Isotone—one of the subdivisions of nuclides used by the physicist; nuclides with the same neutron number but different proton numbers are isotonic brothers.

Isotope—one of the subdivisions of nuclides used by the physicist; nuclides with the same proton number but different neutron numbers are isotopic brothers.

Neutron—one of the basic units of the atom; characterized by the absence of an electrical charge.

Nuclide—the basic physical unit of matter; consists of a dense package of neutrons and protons.

Radionuclide—a combination of neutrons and protons destined to undergo spontaneous transformation through the emission of an alpha particle or an electron (positive or negative in charge) or through fission.

Terminology describing nuclear transformations

Alpha-decay—radiodecay via the emission of the nucleus of a helium atom (that is, a package of two protons and two neutrons).

Beta-decay—radiodecay via the emission of a negatively charged electron.

Electron capture—radiodecay via the entry of an external electron into the nucleus.

Fission—the breakup of a nucleus into two major fragments; in some cases the breakup is spontaneous, in others it is induced by the impact of a neutron.

Fusion—the merger of two nuclides into a single entity.

Half-life—the time required for half of the radionuclides of a given type to undergo radioactive decay.

Mean-life—the average lifetime of radionuclides of a given type.

Neutron capture—the incorporation by a nucleus of a passing neutron.

Positron decay—radiodecay via the emission of a positively charged electron.

p-process—the buildup of heavy nuclides during a proton bombardment.

Radioactive decay—the spontaneous transformation of a radionuclide through particle emission (or capture) or through fission.

r-process—the buildup of heavy nuclides during an intense neutron bombardment.

s-process—the buildup of heavy nuclides during a leisurely neutron bombardment.

Properties of matter

Density—the ratio of an object's mass to its volume.

Mass—a measure of the number of neutrons and protons contained by an object.

Volume—a measure of the space occupied by an object.

Weight—a quantity depending both on the mass of an object and on the gravitational attraction of the planet (or moon) on which the object rests.

Important minerals found on the Earth's surface

Albite—one of the feldspar group consisting mainly of the elements sodium, aluminum, silicon, and oxygen; a primary constituent of granites.

Calcite—a mineral consisting of the elements calcium, carbon, and oxygen; a primary constituent of limestones and marbles.

Olivine—a mineral consisting of the elements iron, magnesium, silicon, and oxygen; a primary constituent of basalts.

Orthoclase—one of the feldspar group consisting mainly of the elements potassium, aluminum, silicon, and oxygen; a primary constituent of granites.

Plagioclase—one of the feldspar group consisting mainly of the elements calcium, aluminum, silicon, and oxygen; a primary constituent of basalts.

Pyroxen—a mineral consisting mainly of the elements iron, magnesium, silicon, and oxygen; a primary constituent of the basalts.

Quartz—a mineral consisting of the elements silicon and oxygen; a primary constituent of granites and of sandstones.

Important rocks found at the Earth's surface

Achondrites—stony meteorites devoid of chondrules.

Basalt—an igneous rock that makes up much of the Earth's continental crust; consists mainly of the minerals olivine, pyroxene, and plagioclase.

Carbonaceous chondrites—a type of chondrite containing many mineral phases which would be destroyed by even moderate heating.

Chondrites—stony meteorites which contain chondrules.

Dolomite—a sedimentary rock consisting primarily of a carbonate mineral (akin to calcite) in which half of the metal atoms are magnesium and half are calcium.

Granite—an igneous rock that makes up much of the Earth's continental crust; consists mainly of the minerals quartz and feldspar.

Igneous rock—a rock that forms when a silicate liquid (i.e., magma) cools and crystallizes.

Iron meteorites—meteorites consisting of alloys of nickel and iron metal.

Limestone—a sedimentary rock consisting primarily of the mineral calcite; most often this calcite was originally manufactured by marine organisms.

Marble—a common metamorphic rock formed when limestones are recrystallized.

Metamorphic rock—a rock that is formed by the recrystallization under heat and pressure of a precursor rock which has become buried deep in the Earth's crust.

Meteorites—rocks which broke free during collisions between asteroids and which subsequently fell upon the Earth.

Sandstone—a coarse-grained rock formed when beach, dune, or river sands become lithified; consists largely of the mineral quartz.

Schist—a common metamorphic rock formed when shales are recrystallized.

Sedimentary rock—a rock formed when material deposited from water, ice, or air becomes cemented into a coherent unit.

Shale—a fine-grained sedimentary rock that originates as mud at the bottom of a standing body of water; consists largely of clay minerals from soils.

Constituents of the Earth's ocean and atmosphere

Carbon dioxide—a trace gas present in the atmosphere and dissolved in the ocean; because of its strong greenhouse effect it plays an important role in climate; chemical formula CO_2.

Hydrogen—the dominant chemical compound in the early solar nebula; now nearly absent from the atmosphere; chemical formula H_2.

Methane—a trace constituent of the Earth's atmosphere with a significant greenhouse capacity; chemical formula CH_4.

Nitrogen—a major constituent of the Earth's atmosphere also present in dissolved form in the ocean; the source of nitrogen fertilizer; chemical formula N_2.

Oxygen—a major constituent of the Earth's atmosphere also present in dissolved form in the ocean; necessary to animal life; chemical formula O_2.

Sea salt—the major ionic species dissolved in the seas are sodium, $Na+$; potassium, $K+$; magnesium, $Mg{++}$; calcium, $Ca{++}$; chloride, Cl^-; and sulfate, $SO_4^=$; important minor ingredients of sea salt are bicarbonate, HCO_3^-; carbonate, $CO_3^=$; nitrate, NO_3^-; and phosphate, $H_3PO_4^-$.

Water—the main ingredient of the ocean and an important greenhouse constituent of the atmosphere; chemical formula H_2O.

Important geologic resources

Coal—carbon-rich residues formed by the heating of plant matter originally deposited as peat in a bog.

Hydrothermal vents—points along spreading axes where seawater circulating through newly created hot basalts spews back into the sea.

Natural gas—methane produced by the thermal "cracking" of molecules in the organic matter contained in sediments.

Oil—liquid hydrocarbons formed by the thermal "cracking" of molecules in the organic matter contained in sediments.

Ore deposits—concentrations of individual metals sufficiently rich to merit commercial recovery.

Salt domes—plugs of solid salt which push up through overlying sediments.

Important geologic structures

Core—a major unit of the Earth's interior consisting of molten iron and nickel; it underlies the Earth's mantle.

Crust—a chemically distinct unit overlying the Earth's mantle.

Lava flow—igneous rock formed by silicate liquids pouring forth from a volcanic vent.

Mantle—a major unit of the Earth's interior consisting of solid silicate minerals; it overlies the Earth's core.

Plates—large sections of the Earth's crust which slide across the underlying mantle as distinct units.

Spreading axes—linear features created where two plates are being carried apart.

Subduction zones—linear features created where one plate is being driven beneath another.

Volcanoes—points on the Earth's surface where liquid silicate (i.e., magma) created in the Earth's interior oozes to the surface.

Astronomical units of matter

Asteroids—small rocky objects that orbit our Sun between Mars and Jupiter.

Binary—a pair of stars that orbit one another.

Black hole—an ultradense remnant of a large star whose nuclear fuel supply has run out.

Comets—small icy objects that orbit our Sun beyond the planet Pluto.

Galaxy—groupings of stars and clouds of gas which constitute the basic unit of matter in the universe.

Moon—an object that orbits a planet.

Nebula—bright or dark patches in the sky caused by clouds of dust and gas or by tight star clusters.

Planet—an object that orbits a central star; distinguished from a star in that it is not sufficiently massive to ignite a nuclear fire.

Red giant—a very large star which burns fast and hot until its fuel is exhausted at which time it undergoes an explosive death.

Star—the basic unit of matter within galaxies; distinguished from other objects (i.e., planets, moons, etc.) by the nuclear fires that burn in their interiors.

Universe—the totality of the matter and radiation in the vast reaches of space observable by the astronomer.

White dwarf—a former star of moderate size that has exhausted its nuclear fuel supply.

Important astronomical phenomena

Big bang—the name given to the explosive origin of our universe.

Three-degree radiation—a homogeneous background glow in the universe thought to be a remnant of the light given off as part of the sequence of events associated with the big bang.

Supernova—the explosive demise of certain large stars.

Figure credits

Pg. 6, Chapter frontispiece. Palomar Observatory 200-inch-telescope photograph of spiral galaxy 29NGC 5194 in Canes Venatici (Copyright © California Institute of Technology).

Pg. 15, Figure 1-2. Red shift—distance relationship. Courtesy of California Institute of Technology.

Pg. 30, Chapter frontispiece. Hale Observatory 200-inch-telescope photograph of Crab Nebula in Taurens U (7NGC-1952), taken in red light. Courtesy Hale Observatories.

Pg. 52, Figure 2-9. Schmidt 48-inch-telescope photographs taken before and after a supernova event in NGC 5253. Courtesy of Hale Observatories.

Pg. 66, Chapter frontispiece. Picture of a circumstellar disk around the star Beta Pictoris taken from the Los Campanos Observatory near La Serena, Chile using a charge-coupled device. The disk is likely composed of the materials from which our own planets formed. The images were processed by Bradford Smith of the University of Arizona and Richard Terrile of the Jet Propulsion Laboratory. Courtesy of Jet Propulsion Laboratory.

Pg. 77, Figure 3-1. Photomicrograph of Brownfield H-3 chondritic meteorite. Courtesy Carlton Moore, Arizona State University.

Pg. 81, Figure 3-4. Polished section of the Bagdad A_2 iron meteorite. Courtesy of Carlton Moore, Arizona State University.

Pg. 91, Figure 3-6. Based on a diagram in A.E. Ringwood, *Origin of the Earth and Moon*, 1979.

Pg. 96, Chapter frontispiece. Tieschitz H-3 chondrule. Courtesy of Martin Prinz, American Museum of Natural History, New York, N.Y.

Pg. 128, Chapter frontispiece. Polished section of Springwater stony iron meteorite. Courtesy of Carlton Moore, Arizona State University.

Pg. 159, Figure 5-12. Map showing age distribution of the seafloor basaltic crust. Courtesy of R.L. Larson, University of Rhode Island and W. Pitman, Lamont-Doherty Geological Observatory (Copyright © 1985).

Pg. 163, Chapter frontispiece. Arend-Roland Comet 297C, photographed in April 1957 using a 48-inch telescope. Courtesy Hale Observatories.

Pg. 167, Figure 6-1. Information about the moons of the major planets. Adapted from a diagram in *The Solar System* by John Wood, 1979. Densities updated.

Pg. 169, Figure 6-2. Voyager 1 image of Saturn's moon Rhea and Voyager 2 image of Saturn's moon Titania. Courtesy of Jet Propulsion Laboratory.

Pg. 170, Figure 6-3. Viking Orbiter image of Mars' moon Phobos. Courtesy Jet Propulsion Laboratory.

Pg. 173, Figure 6-5. Based on a diagram in A.E. Ringwood, *Origin of the Earth and Moon.*

Pg. 176, Figure 6-7. X-ray photograph of coral slab from the Red Sea. Courtesy Richard Cember, Lamont-Doherty Geological Observatory.

Pg. 180, Figure 6-8. Based on information supplied to the author by Robert Clayton of the University of Chicago.

Pgs. 183 and 184, Figures 6-9 and 6-10. Digitized images of the front and back side of the Moon based on an airbrush map. Courtesy of Jody Swann, U.S. Geological Survey, Flagstaff, Arizona.

Pg. 185, Figure 6-11. Viking 1 photograph of the Lunae Planum region of Mars. The crater shown in this picture is about 30 kilometers in diameter. A dry river channel is also seen. Courtesy Jet Propulsion Laboratory.

Pg. 186, Figure 6-12. Photomosaic taken by Mariner 10 of the south pole of Mercury. Courtesy Jet Propulsion Laboratory.

Pg. 187, Figure 6-13. Aerial photograph of the Arizona Meteorite Crater. Courtesy of John Sewall Shelton, La Jolla, California.

Pg. 188, Figure 6-14. Photograph of a number of tectites. Courtesy of Martin Prinz, American Museum of Natural History, New York, N.Y.

Pg. 196, Chapter frontispiece. Aerial photograph of the Barnard Glacier. Courtesy of Robert Sharp, California Institute of Technology and the University of Oregon Press, Eugene, Oregon.

Pg. 216, Figure 7-6. Maps showing ice extent today and during peak glacial time. Courtesy George Kukla, Lamont-Doherty Geological Observatory.

Pg. 225, Figure 7-14. Climate spectra. Courtesy of John Imbrie of Brown University.

Pg. 228, Chapter frontispiece. Pyrite crystals. Courtesy of Ross M. Horowitz Photography, New York, N.Y.

Pg. 236, Figure 8-2. Precambrian fossil animals. Courtesy of R. Sprigg.

Pg. 258, Chapter frontispiece. Photograph of thermonuclear explosion carried out in the Marshall Islands in the fall of 1952. Courtesy U.S. Navy and Livermore National Laboratory.

Index